BestMasters

Springer awards „BestMasters" to the best master's theses which have been completed at renowned universities in Germany, Austria, and Switzerland.

The studies received highest marks and were recommended for publication by supervisors. They address current issues from various fields of research in natural sciences, psychology, technology, and economics.

The series addresses practitioners as well as scientists and, in particular, offers guidance for early stage researchers.

Michael Schenk

Studies with a Liquid Argon Time Projection Chamber

Addressing Technological Challenges of Large-Scale Detectors

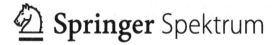 **Springer** Spektrum

Michael Schenk
Bern, Switzerland

BestMasters
ISBN 978-3-658-09429-4   ISBN 978-3-658-09430-0 (eBook)
DOI 10.1007/978-3-658-09430-0

Library of Congress Control Number: 2015935658

Springer Spektrum

Printed on acid-free paper

Springer Spektrum is a brand of Springer Fachmedien Wiesbaden
Springer Fachmedien Wiesbaden is part of Springer Science+Business Media
(www.springer.com)

Preface

This master thesis is the outcome of 16 months of research and development work on the ARGONTUBE project at the Laboratory for High Energy Physics and Albert Einstein Center for Fundamental Physics (LHEP-AEC) of the University of Bern. This project has been initiated specifically to evaluate certain aspects of the feasibility of large and massive, i.e. kiloton-scale, liquid argon time projection chambers (LArTPC) and to address the associated technological challenges. LArTPCs are a promising particle detector type offering the detection capabilities required amongst others by neutrino and astrophysics experiments.

For my master thesis, I have become fully involved into this fascinating project and have taken the opportunity to make my own contributions to the development of this particle detector technology. I have participated in five measurement campaigns where a large number of UV laser induced events as well as cosmic muon tracks have been acquired with a novel cryogenic read-out. These data allowed the extraction of interesting results, relevant for the ARGONTUBE project as well as for future large LArTPCs in general. In between the runs, I was involved in doing technical work, modified and prepared the detector and associated systems. I have gained experience in the fields of vacuum technique, cryogenics, electronics and programming.

Apart from a general introduction to the LArTPC technology, this master thesis reports on the ARGONTUBE detector upgrades that have been realized as well as the measurements and studies that have been carried out.

During my work on the ARGONTUBE project, I have been supported by a number of persons to whom I wish to express my sincere gratitude.

First of all, Prof. Dr. Antonio Ereditato who initiated this master thesis and allowed me to be part of such an interesting research project.

PD Dr. Igor Kreslo and Prof. Dr. Michele Weber, who have been excellent mentors and advisors during my involvement in the ARGON-TUBE project. I thank you for sharing your valuable knowledge and experience with me. I enjoyed the freedom and responsibility I was given to carry out my studies.

Dr. Marcel Zeller and Christoph Rudolf von Rohr for teaching me everything about ARGONTUBE. I have highly appreciated your support and ideas. Furthermore, Prof. Dr. Štefan Jánoš, Sébastien Delaquis and Dr. Akitaka Ariga for their precious technical advice on related topics.

Igor Kreslo, Michele Weber and Thomas Strauss for proof-reading this document and giving numerous valuable inputs to my thesis.

My fellow students and friends Sabina Joos, Matthias Lüthi and Marcel Häberli with whom I spent most of the time during my master studies. A big thank you to all of you.

Finally, my dear family and friends for supporting me in all the ways possible throughout my studies at the University of Bern.

Michael Schenk

Contents

List of Figures

Introduction

The field of neutrino physics has an ever increasing demand for more sophisticated particle detectors. It was initiated in 1930 by Wolfgang Pauli [1], who proposed a new, electrically neutral spin 1/2 particle, later called the 'neutrino', to explain the continuous energy spectrum of the nuclear β-decay electron. In 1934, Enrico Fermi published his successful theory on the β-decay including the neutrino as one of the decay products [2]. Its experimental detection, however, turned out to be very challenging and it took more than 20 years until it was unambiguously discovered (they were antineutrinos in fact) by Frederick Reines and Clyde L. Cowan in 1956 [3]. Since then, the neutrino has been studied in various experiments which revealed at least some of its peculiar properties, the major one being the oscillations between the three different flavours electron, muon and tau, originally predicted by Bruno Pontecorvo in 1957 and later confirmed by numerous experiments (see for instance [4] and references therein).

Even though our knowledge about the neutrino has steadily grown over the past decades, many observations are still puzzling. To obtain the sensitivity required to answer the open questions related to neutrino physics, a massive (order of 10 kt to 100 kt) and high precision detector is required. It must feature fine grained 3D imaging, calorimetric and superior particle identification capabilities to allow efficient reconstruction of neutrino interactions. One detector type which fulfills these requirements is the liquid argon time projection chamber (LArTPC). A 600 t LArTPC has already been built and successfully operated by the ICARUS collaboration at the National Laboratory in Gran Sasso (LNGS) [5]. However, the technological challenges arising with ever larger detector masses are yet to be studied in detail. A major one is the long charge drift distance of 10 m to

20 m which requires amongst others voltages of up to 1 MV to 2 MV, oxygen equivalent impurities on a level of less than 0.1 ppb (10^{-10}) dissolved in the target liquid argon volume and a low-noise charge read-out. These aspects can be explored reliably only with a long drift distance LArTPC prototype which is why the ARGONTUBE research and development project has been initiated at the Laboratory for High Energy Physics and Albert Einstein Center for Fundamental Physics (LHEP-AEC) of the University of Bern [6–8]. The ARGONTUBE LArTPC has a record-setting 5 m long charge drift distance. It serves as a powerful instrument to study existing technological challenges for future kiloton-scale LArTPCs and to evaluate and develop possible solutions.

The main goal of this thesis is to acquire high-quality cosmic muon and UV laser induced event samples with ARGONTUBE and to extract new results, relevant for future studies with ARGONTUBE and for the operation of large LArTPCs in general. To achieve this goal, a number of intermediate steps were necessary, including modifications and upgrades of the detector, development of calibration and data analysis methods as well as the implementation of software tools. After introducing the main concepts of a LArTPC in Chapter 1, the ARGONTUBE detector, along with the systems that are necessary for its operation, are explained in Chapter 2. This includes the description of modifications and upgrades of the detector that have been realized and tested within the scope of this master thesis. For detector operation, a new regeneration system for argon purifiers was designed, built and tested. The results are discussed in Chapter 3. Cryogenic preamplifiers, developed particularly for large LArTPCs, replaced the 'warm' read-out scheme in ARGONTUBE and allowed the acquisition of cosmic muon and UV laser induced events with a significantly higher signal-to-noise ratio than formerly achieved. To extract meaningful results from these events, several intermediate steps had to be taken. The electric field disuniformities in ARGONTUBE, originating from the high voltage generator, were studied by means of UV laser induced events and a model for the electric field in the sensitive detector volume was determined. It allowed decoupling of the longitudinal from the transverse field component and yielded an estimate for the strength of the parasitic transverse component in ARGONTUBE. The methods are explained in Chapter 4. To process the muon events, a track finder employing the Hough transform was implemented. Compute-intensive parts of the code were outsourced to a graphics processing unit, a massively parallel computing architecture. This led to a significant speed-up as discussed in Chapter 5. By using the track finder and the model for the electric field, the muon energy loss per unit track length in liquid argon was measured, both for the 'warm' and the cryogenic read-outs. Moreover, a method for the momentum estimation by multiple Coulomb scattering [9] was implemented and applied to ARGONTUBE muon events. These studies are presented in Chapter 6. A new approach to determine the amount of impurities in ARGONTUBE with the UV laser system is introduced in Chapter 7 and verified by means of cosmic muon tracks. Finally, a study of the longitudinal diffusion coefficient in liquid argon for electric field strengths ranging from 120 ± 49 (RMS) V/cm to 210 ± 84 (RMS) V/cm is presented in the same chapter.

1 The liquid argon time projection chamber

The time projection chamber (TPC) was invented by David Nygren in 1974 at the Lawrence Berkeley National Laboratory (LBNL) in California, USA [10]. It followed the invention of the multi-wire proportional chamber (MWPC) which was made in 1968 [11] by Georges Charpak who was awarded the Nobel Prize in Physics in 1992 for this reason. The development of these devices marked a revolutionary step in the detection of ionizing radiation. For the first time it was possible to read out electronically a detector with tracking capabilities as superior as those of a bubble chamber. Originally, TPCs were typically filled with noble gases where argon was the first choice in most cases. In 1977, Carlo Rubbia [12] proposed the use of liquefied argon as a target medium in TPCs for reasons that will be explained in this chapter. The introduction on the LArTPC working principle is followed by a more detailed discussion of the main aspects and processes associated with this detector technology.

1.1 Working principle

A TPC consists of two parallel planes, cathode and anode, separated by the drift gap that can range from a few mm to several m depending on the field of application of the device. While the anode is connected to ground, the cathode is biased to a high negative electric potential to set up an electric field within the detector active volume. The field strength is typically of the order of several $100\,\mathrm{V/cm}$ for the largest drift gaps and up to a few $10\,\mathrm{kV/cm}$ for the smallest ones. To guarantee uniformity of the field across the entire sensitive detector

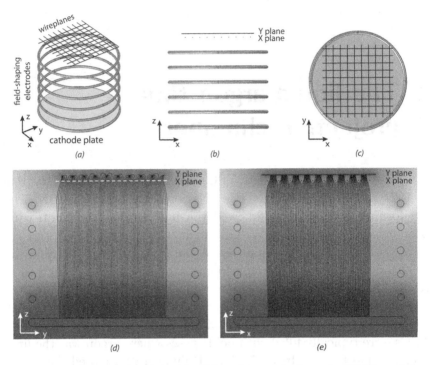

Figure 1.1: Perspective (a), side (b) and top (c) view of a TPC model. Images (d) and (e) illustrate the electric field lines (in black) and electric potential obtained with a simulation in the volume central region for the two projections y-z and x-z respectively.

volume, equally spaced field-shaping electrodes are installed in between the anode and the cathode, properly biased in their electric potentials with respect to one another. The anode is given by the sensing plane which is where the detector output signals are formed and registered by means of a spatially segmented read-out, consisting of two or more wireplanes oriented along different directions. A wireplane is an array (grid) of thin conductive wires each of which represents one detector read-out channel. Instead of wireplanes, pad electrodes can be used (pixel read-out). A drawing of a cylindric TPC with a two-wireplane read-out is shown in Figure 1.1 in a perspective (a), side (b) and top (c) view. The ring electrodes that are equally distributed along the z axis form the field cage. The conductive plate at the very bottom is the cathode. Figure 1.1 (d) and (e) show the results of a COMSOL[1] finite element method (FEM) simulation of the electric field within the sensitive volume of the chamber in the two cut-views y-z and x-z. The field is highly uniform for the given field cage geometry. When immersing the chamber in an appropriate medium, such as liquid argon, it works as an excellent, fully homogeneous particle detector. Liquefied argon plays the role of both, sensitive and target volume. Since it is a cryogenic medium, the chamber is contained in a cryostat.

An ionizing particle traversing a LArTPC creates pairs of positively charged argon ions Ar^+ and quasifree electrons e^- along its path (ionization track). Instead of being ionized, an argon atom may be raised to an excited state, which eventually leads to the emission of argon scintillation light during its de-excitation. The situation is illustrated in Figure 1.2 (a) for a passing-through muon μ. Right after creation, an electric field dependent amount of the electron-ion pairs recombines, resulting in the emission of more scintillation light. Liquid argon is transparent to its own scintillation light. It can thus be measured by photodetectors to use it as a trigger, to provide the precise event time stamp t_0 and to gain additional information about an event to facilitate its reconstruction. The residual electrons and argon cations, left after recombination, are separated by the presence of the electric field forcing them to drift towards the anode or the cathode respectively. Their movements are described by phases of acceleration, interrupted by collisions with the surrounding atoms of the medium. This can be modelled by a finite average electron or ion drift speed. Compared to the argon ions, electrons have a mobility which is higher by orders of magnitude and, at a given electric field strength, they have much higher drift speeds. In terms of overall detector performance, a fast charge transport is favourable which is why the electrons rather than the ions are read out to obtain

---

[1]www.comsol.com

information about the ionization track[2]. Due to the high uniformity of the electric field, the integrity of the original ionization track is kept during the drifting process (c). Any field disuniformities would cause distortions of the track and strongly affect the detector spatial resolution.

During the drift, two physics processes have an impact on the detector performance. Firstly, the drifting electrons are subject to longitudinal (along the drift direction) and transverse (perpendicular to the drift direction) diffusion, meaning that they do not strictly keep their positions relative to one another, but rather they disperse. This limits the spatial resolution of the device. Secondly, electronegative impurities, such as oxygen and water molecules dissolved in the sensitive liquid argon volume, tend to attach electrons and thus reduce the amount of charge drifting towards the read-out (c). Consequently, impurities diminish the detector output signals and hence one is interested in keeping their concentration in the device as low as possible. At the sensing plane, the electrons are registered by the x-y segmented read-out (d) and the signals produced are subsequently amplified electronically. In addition to measuring the x and y spatial coordinates, the electron arrival times are recorded. Having knowledge of the event time stamp t_0, for instance from the measurement of the scintillation light, the actual drift times t_d can be calculated from the arrival times. Since t_d is proportional to the drift distance, it is possible to obtain the third spatial coordinates z of the ionization track to allow full 3D tracking. After doing a thorough detector calibration and applying appropriate corrections for attachment and recombination losses, the number of electrons collected at the read-out plane yields calorimetric information about an event.

1.2 Liquefied argon as a detection medium

Liquefied noble gases are very attractive particle detector media as they uniquely combine a number of properties. The main elements of

---

[2]There are also attempts on reading out ions [13].

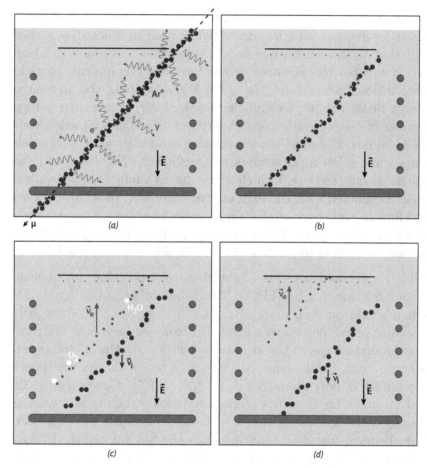

Figure 1.2: (a) A muon μ passes through the detector volume and creates an ionization track (Ar^+ - e^- pairs) and leads to the
emission of UV scintillation light γ. (b) A certain fraction
of the electron-ion pairs have recombined (100 % in regions
of zero electric field), the residual ones are separated by the
applied electric field. (c) Electrons drift towards the read-out
(i.e. upwards in the figure), the argon cations towards the
cathode plate (i.e. downwards). The integrity of the original
ionization track is kept thanks to a uniform drift field. A
certain number of the drifting electrons are subject to attachment by electronegative impurities. (c)/(d) The electrons
are registered by the segmented read-out together with their
arrival times.

interest are argon, xenon and krypton. They provide a high stopping power for ionizing radiation, described by their high densities ρ, short radiation X_0 and nuclear interaction λ_I lengths. This results in a large energy loss per unit distance dE/dx for traversing ionizing particles and, taking into account the value W_i, describing the amount of energy needed to produce an electron-ion pair, in a relatively large number of quasifree electrons. Apart from the high ionization yield, noble liquids offer excellent scintillation properties with a high light output and a low mass attenuation coefficient. The atoms of noble liquids do not tend to attach electrons due to a fully occupied valence shell. Combined with the high electron mobility, these media allow an efficient transport of electrons across long distances, an aspect of major relevance for TPCs. Moreover, the technology to strongly reduce the level of impurities in noble liquids is readily available.

Table 1.1 summarizes some of the characteristic values for liquefied argon, relevant for LArTPCs. Note that other noble liquids like xenon would be even more attractive in terms of ionization yield, stopping power or temperature [14]. However, argon is the third-most abundant gas in the atmosphere with a relative abundance of nearly 1 %vol., while xenon is extremely rare with an atmospheric abundance of only about $8.7 \cdot 10^{-6}$ %vol. [15]. Consequently, the cost for argon are by orders of magnitude lower than those for xenon, making it the best candidate for a kiloton-scale liquid noble gas TPC. The chemical inertness resulting from the full valence shell (argòs = 'inactive') is advantageous in terms of safety as there is no fire hazard. However, there is the danger of suffocation for instance for argon gas as it has a higher density than air. This must be taken into account for a large LArTPC which would be built and operated in an underground laboratory. The major drawback in using liquefied noble gases for particle detectors is the need for a cryogenic environment. It complicates the handling and the operation of such devices. The aspect of radiopurity, that is the amount of radioactive isotopes that are present in the detector medium, has to be addressed particularly when working in low or zero background experiments. Despite the drawbacks, the abovementioned noble liquids, particularly argon for large mass detectors, serve as excellent detector media.

Table 1.1: A selection of liquid argon properties. Values are taken from [15–17] and references therein. η_{atm} is quoted for a standard U.S. atmosphere 1976 [15]. T_{BP} and T_{MP} correspond to a pressure of 1 atm. W_i, W_s and $\langle dE/dx \rangle$ are valid for minimum ionizing particles. μ_0^e, μ_0^i and v_e, v_i are specified at T_{BP} and respectively $E = 0\,\text{kV/cm}$ (zero field mobilities) and $E = 0.5\,\text{kV/cm}$.

Quantity	Symbol	Value	
Atomic number	Z	18	
Molar mass	M	39.948(1)	g/mol
Atmospheric abundance	η_{atm}	0.934	%vol.
Boiling point (BP)	T_{BP}	87.26	K
Melting point (MP)	T_{MP}	83.8(3)	K
Density at normal BP	ρ_{BP}	1396(1)	kg/m^3
Dielectric constant	ϵ_r	1.505(3)	
Ionization energy	I	13.84	eV
W-value for ionization	W_i	23.6(3)	eV/pair
W-value for scintillation	W_s	19.5(10)	eV/photon
Radiation length	X_0	14.0	cm
Nuclear interaction length	λ_I	85.7	cm
Mean specific energy loss	$\langle dE/dx \rangle$	2.12	MeV/cm
Scintillation emission peak	λ_s	128	nm
e^- drift velocity	v_e	1.60(2)	mm/μs
e^- mobility	μ_0^e	518(2)	cm^2/(Vs)
Ar$^+$ drift velocity	v_i	$8.0(4) \cdot 10^{-6}$	mm/μs
Ar$^+$ mobility	μ_0^i	$6.0 \cdot 10^{-4}$	cm^2/(Vs)

1.3 Energy dissipation in liquefied argon

Depending on the type of radiation, the energy of a particle traversing a medium is dissipated in different ways [18–21]. Hadrons, such as the neutron, lose their energy by the short-range strong interactions with nuclei of the medium. In case they are electrically charged, the hadrons also take part in electromagnetic interactions described further below in the text. Concerning neutrinos and antineutrinos crossing the detector, they can be absorbed by the nuclei of the medium in weak interactions.

Photons or γ-rays lose their energy via the processes of photoelectric absorption, Compton scattering and pair production. In the process of photoelectric absorption, an incoming photon fully transfers its energy E_γ to one of the shell electrons of an atom. The electron is removed from the shell and its final state kinetic energy amounts to $E_{\text{kin}} = E_\gamma - E_B$, where E_B denotes its initial atomic binding energy. Compton scattering describes the process where the incoming photon scatters off an electron. During such an interaction, only part of E_γ is transferred to the electron and the photon is not absorbed but its wavelength is increased and its initial direction of flight changes. The process of pair production may only occur if $E_\gamma > 2m_e$ with m_e being the mass of an electron or a positron respectively. The photon produces an electron-positron pair in the presence of a so called spectator nucleus needed to fulfill the momentum conservation law. Which of the three processes dominates depends mainly on the target medium (proton number Z) and on E_γ. Roughly, the interaction cross-section dependences of photoelectric absorption, Compton scattering and pair production are respectively $\sigma_A \sim Z^5/E_\gamma^{7/2}$, $\sigma_{CS} \sim Z/E_\gamma$ and $\sigma_{PP} \sim Z^2 \ln (2E_\gamma)$, the latter with a threshold of $E_\gamma > 2m_e$ [19].

For electrically charged particles, electromagnetic interactions with the atomic shell lead to ionization, excitation and transfer of heat [19, 20]. The latter becomes the dominant process below a certain energy threshold when only elastic collisions with the medium occur and ionization and excitation are no more possible. Heat transfer losses are hence not measured in a TPC. Apart from the three scattering interactions, a relativistic charged particle deposits a fraction of its energy by emitting radiation, such as producing bremsstrahlung γ-rays. The energy scale where this process starts to dominate is strongly dependent on the particle mass. In the process of ionization, enough energy is transferred from the ionizing particle to one of the shell electrons of an atom to overcome its binding energy

$$\text{Ar} + \chi \rightarrow \text{Ar}^+ + e^- + \chi', \tag{1.1}$$

where χ and χ' denote the ionizing particle before and after the interaction respectively and Ar represents the argon atom. Until it thermalizes, the resulting electron may have enough energy E_e for further ionization or excitation of atoms. If $E_e \gg I$, where I is the ionization potential, the secondary electron is also called a δ-ray [21]. If the energy transferred to the shell electron is smaller than the ionization potential of the medium, the atom may end up in a highly excited state Ar^{**} [22]

$$\text{Ar} + \chi \rightarrow \text{Ar}^{**} + \chi'. \tag{1.2}$$

The shell electron to which the energy was transferred is lifted to a higher energy state while remaining bound to the atom. These excited states are short-lived. By collisions with surrounding argon atoms, de-excitation to the first excited state Ar^* occurs, either through non-radiative energy relaxation (heat), or by the emission of one or several vacuum ultraviolet (VUV) photons

$$\text{Ar} + \text{Ar}^{**} \rightarrow \text{Ar} + \text{Ar}^* + \text{heat}, \tag{1.3}$$
$$\text{Ar} + \text{Ar}^{**} \rightarrow \text{Ar} + \text{Ar}^* + \gamma. \tag{1.4}$$

The de-excitation of the Ar^* eventually results in the emission of scintillation light. The mechanisms are explained in Section 1.5.

The amount of charge produced by ionization is directly related to the energy deposited in the argon. The conversion factor is given by the value W_i, but recombination and electron attachment losses must be taken into account. Since the energy deposited by the ionizing particle does not completely go into ionization, but also into excitation and heat, $W_i > I$. The energy deposited per unit track length dE/dx, also

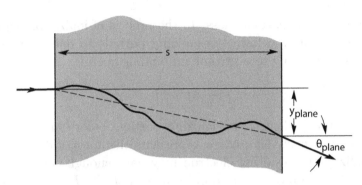

Figure 1.3: A charged particle entering a layer of thickness s of a medium is deflected by many small angle scatterings (multiple
Coulomb scattering MCS) and leaves the material with a
displacement y_{plane} and under a certain angle θ_{plane} (plane-
projected values) [21].

known as the linear energy transfer (LET), is an important quantity
for particle identification and can be precisely measured in LArTPCs
thanks to their excellent 3D tracking and calorimetric capabilities.
For moderately relativistic heavy ($m \geq m_\mu$) charged particles, the
Bethe-Bloch formula provides a good description of dE/dx [21].

The precise tracking capabilities of a LArTPC allow the measurement
of a particle's momentum even if its ionization track is not fully
contained in the detector volume. It can be done by analysis of the
many small angle scatterings (multiple Coulomb scattering MCS)
that the particle undergoes when traversing the medium [9]. The
process is illustrated in Figure 1.3. A particle enters a layer of
thickness s of a certain material and is scattered multiple times
through electromagnetic interactions (and strong interactions in the
case of hadronic particles) mainly by small angles. The particle exits
the layer with an offset y_{plane} with respect to the original position and
under an angle θ_{plane} (plane-projected values) relative to its initial
direction of flight. The angular distribution of θ_{plane} is well described
by Molière theory [21]. The central part (98 %) is Gaussian while the
tails are extended, i.e. larger than Gaussian tails. They correspond to

large angle scatterings which are explained by Rutherford scattering. The RMS width of the central part is given by [21]

$$\theta_0 = \frac{13.6\,\text{MeV}}{\beta p c} z \sqrt{\frac{s}{X_0}} \left[1 + 0.038 \ln \left(\frac{s}{X_0} \right) \right], \qquad (1.5)$$

where βc, p and z are the velocity, momentum and the electric charge of the incident particle and X_0 denotes the radiation length of the traversed medium. By analysis of the root-mean-square (RMS) values of θ_{plane} along the particle track for different layer thicknesses s, one can use Equation 1.5 to find an estimate for βp. If, in addition, the particle mass is known, p can be calculated.

Straight ionization tracks in liquid argon can be induced by a pulsed and intense UV laser beam. The ionization energy of an argon atom ($I = 13.84\,\text{eV}$, see Tab. 1.1) would require a wavelength $\leq 89.8\,\text{nm}$ to generate electron-ion pairs by single photon absorption. Lasers producing light of these wavelengths are not commercially available. Additionally, the transmittance in air would be very small causing the need for an evacuated beam path. However, UV light of $266\,\text{nm}$ ($4.67\,\text{eV}$) can induce ionization if several photons transfer their energy 'simultaneously' to one and the same shell electron of an argon atom – a process that is called multi-photon absorption [23–25]. The photon flux must be high enough for this process to be efficient. For gaseous argon, an atom has four distinct excited states at around $12\,\text{eV}$, and the ionization potential is about $15.8\,\text{eV}$. As soon as the gas condenses to a liquid, the discrete excited states are broadened to bands and are lowered in energy to $9.63\,\text{eV}$ and $9.8\,\text{eV}$ respectively. Even though ionization by multi-photon absorption may occur via laser-induced virtual states that are not necessarily eigenstates of the argon atoms, it is beneficial if there are real excited atomic states close to the virtual states (quasi-resonant ionization). The reason for that are the lifetimes of $10^{-16}\,\text{s}$ and $10^{-8}\,\text{s}$ of the laser-induced virtual states and the real excited states respectively [24]. Whenever two photons of energy $E_\gamma = 4.67\,\text{eV}$ are 'simultaneously' absorbed by one and the same shell electron, the argon atom is raised to an excited state. If there is a third photon of energy E_γ absorbed by the same electron within the timescale of the lifetime of the excited state, the atom may be ionized. More detailed discussions and measurements concerning this process are given in [23–25].

1.4 Recombination mechanisms

A fraction of the electron-ion pairs created during the ionization process in argon are not separated fast enough by the applied electric field and thus recombine. A variety of models exist [26–28] to describe these processes, but they are not yet entirely understood. The main idea is that after thermalization through collisions with particles of the surrounding medium (within 1-2 ns [29]), the quasifree electrons created during the ionization process may remain close enough to an ion to be captured again. The fraction of electron-ion pairs that recombine is strongly dependent on the applied electric field. With an increasing field strength, the Coulomb force separating ions from electrons gets larger and hence there is a smaller chance for recombination. The dependence of the recombination fraction on the electric field strength and on the specific energy loss dE/dx of the ionizing particle is described by a saturation curve which is shown in Figure 1.4 for liquefied argon.

In 1938, Onsager proposed that recombination mainly happens through re-attachment of the electron to its parent ion due to the attractive Coulomb force [26]. He assumed that each electron-ion pair can be looked at individually without considering the influence of all the other ones which is why it is called a geminate theory. This would imply that recombination was not dependent on the ionization density. The distance from the cation within which the thermalized electron recombines is called the Onsager radius r_c. It describes the distance at which the potential energy of the Coulomb attraction of the electron to its parent ion is equal to the thermal kinetic energy of the electron

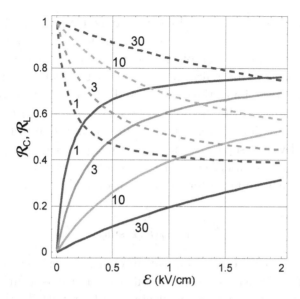

Figure 1.4: The recombination factors for charge (solid lines) and light
(dashed lines) as functions of the electric field strength. R_C
and R_L denote the collected charge and light at the given
electric field divided by respectively the charge collected
at infinite field and the light collected at zero field. The
numbers that label the curves denote the specific energy
loss dE/dx of the particle in units of minimum ionizing
particles [16].

$$\frac{e^2}{4\pi\epsilon_0\epsilon_r r_c} = kT \quad \Leftrightarrow \quad r_c \doteq \frac{e^2}{4\pi\epsilon_0\epsilon_r kT}, \qquad (1.6)$$

where e is the elementary charge, ϵ_0 and ϵ_r are respectively the vacuum
permittivity and the dielectric constant, k is Boltzmann's constant
and T is the temperature. In liquid argon, the Onsager radius is
$r_c = 125\,\text{nm}$.

A different approach was chosen by Jaffé in 1913 [27]. His model is
based on a columnar theory, meaning that recombination is expected
to depend on collective effects, i.e. on the charge density of electrons

and ions within a cylinder along the ionization track. Instead of recombining only with its parent ion, an ionization electron may recombine with any other of the nearby ions. Consequently, the fraction of electron-ion pairs that is subject to recombination would also depend on the ionization density and thus on the type of particle that passes through the medium. The column of electrons and ions evolves as a function of time due to diffusion and due to the applied electric field which separates negative from positive charge carriers. During this process, an electron may be captured by one of the nearby ions and the charge left after recombination is described by

$$Q = \frac{Q_0}{1 + q_0 F(E \sin\phi)},$$ (1.7)

where Q_0 denotes the total ionization charge before recombination, q_0 is the initial density of electron-ion pairs and F is a function that depends amongst others on the electric field strength E, the angle ϕ between electric field and ionization track as well as other parameters that describe diffusion. The fraction of charge that does not recombine is given by $R = Q/Q_0$. The Jaffé model assumes the same drift speed for ions and electrons and their mobilities are regarded as constants.

By calculation of the average ion-ion and the electron-ion distances in liquid argon, one can evaluate the relevance of geminate recombination (Onsager model) compared to the columnar effects described by Jaffé. By knowledge of W_i and dE/dx in liquid argon, the average distance between ions is calculated to be of the order of 10 nm to 50 nm depending on the type of the ionizing particle. By contrast, electrons travel distances of about 10^3 nm before thermalization [30] which is not only much larger than the distance between neighbouring ions, but also about ten times the Onsager radius r_c. Thus, one expects geminate recombination to be disfavoured and that collective effects play an important role during the process. In fact, experimental data indicate that the fraction of electron-ions pairs that recombine depends on the ionization density [30, 31]. The curves shown in Figure 1.4 originate from a reliable fit of the columnar model to experimental data. For higher dE/dx, the density of electron-ion pairs along the track is higher and thus a larger fraction recombines.

Thomas and Imel [28] reformulated the Jaffé columnar equation (see Eq. 1.7) by assuming diffusion and ion mobility to be zero. They model the cylindric column by a box that contains a uniform distribution of charge (box model)

$$Q = Q_0 \frac{1}{\xi} \ln(1 + \xi), \qquad (1.8)$$

where $\xi = \alpha Q_0/E$ with α being a free parameter. For liquid argon, experimental data on recombination are usually fitted using the functional form coming either from the Jaffé model (see Eq. 1.7) or from the box model (see Eq. 1.8). Typically, the Jaffé model is approximated by means of Birks' law [32], which was originally used to describe quenching effects in scintillators. It is applied either in the form

$$Q = A \cdot \frac{Q_0}{1 + k_E/E} \quad \text{or as} \quad Q = A \cdot \frac{Q_0}{1 + k_Q\, dE/dx}, \qquad (1.9)$$

where respectively k_E and k_Q are fit parameters. The normalization parameter A is added to match the experimental data better in the low field region [31]. The parameter values that were found by the ICARUS collaboration [31] from fitting Equation 1.9 to data are

$$A = 0.800 \pm 0.003 \text{ and } k_E = 0.0486 \pm 0.0006\,\text{kV/cm}\frac{\text{g/cm}^2}{\text{MeV}}. \qquad (1.10)$$

These values have been confirmed by [30], which also contains more detailed studies of experimental data in the context of the columnar models.

To obtain calorimetric information for an event in a LArTPC, the collected charge must first be corrected for electron attachment losses caused by impurities, and second for recombination. The latter is done while transforming the charge collected at the TPC read-out ΔQ, corrected for attachment losses, into the equivalent amount of energy ΔE

$$\Delta E = \frac{\Delta Q}{R} W_i \quad \text{or} \quad dE/dx = \frac{dQ/dx}{R} W_i, \quad (1.11)$$

where dQ/dx and dE/dx are respectively the residual charge after recombination and the energy deposited per unit track length. By means of Equation 1.9,

$$dE/dx = \frac{dQ/dx}{A/W_i - k \cdot (dQ/dx)/E}, \quad (1.12)$$

where k is given by k_E normalized to the density of liquid argon $k = k_E/\rho_{Ar}$. dQ/dx is in units of number of electron-ion pairs per unit track length.

Recombination occurs either directly via $Ar^+ + e^- \to Ar^*$, followed by de-excitation of the Ar^* via emission of VUV photons or through non-radiative relaxation (heat), or through a triple collision of the argon cations with surrounding argon atoms. The latter is very efficient (occurs within ~ps) due to the high density of the liquid. It results in the formation of an ionic excimer state Ar_2^+ [33]

$$Ar^+ + 2\,Ar \to Ar_2^+ + Ar. \quad (1.13)$$

A third argon atom has to take part in this interaction to guarantee momentum conservation. The term excimer (short for excited dimer) describes a compound of two atoms (molecules) that is strongly bound only in its excited states. Recombination happens with the ionic excimer Ar_2^+ which leads to a highly excited state Ar^{**} of an argon atom

$$Ar_2^+ + e^- \to Ar^{**} + Ar. \quad (1.14)$$

Again, the highly excited Ar^{**} relaxates to the first excited state Ar^* by the two mechanisms shown in Equations 1.3 and 1.4. The first excited states subsequently de-excite under the emission of light. As a result, one expects the light yield to be higher when a larger fraction of electron-ion pairs recombines and hence, at a given electric field strength, the light yield increases with higher dE/dx. This anti-correlation behaviour between collected charge and light is visible in Figure 1.4.

1.5 Scintillation mechanisms

Figure 1.5 gives a simplified picture of the two major scintillation mechanisms that occur in liquid argon. Further mechanisms are discussed in [22, 34] and references therein. The interactions that take place in a first step are direct excitation by the ionizing radiation (see Eq. 1.2) and recombination of the electron-ion pairs (see Eq. 1.14). In both cases, the intermediate state is given by an argon atom in its first excited state Ar* (see Eqs. 1.3 and 1.4). In triple collisions with surrounding argon atoms, the excited argon atom can change its electronic configuration and form an excimer Ar_2^*. The formation happens on a ps timescale [33]

$$Ar^* + 2\,Ar \rightarrow Ar_2^* + Ar. \tag{1.15}$$

The excimer is either bound in the singlet state $^1\Sigma_u^+$ or in the triplet state $^3\Sigma_u^+$ with different lifetimes after which de-excitation occurs [22, 34]

$$Ar_2^*(^1\Sigma_u^+) \rightarrow Ar_2(^1\Sigma_g^+) + \gamma \rightarrow 2\,Ar + \gamma \tag{1.16}$$

$$Ar_2^*(^3\Sigma_u^+) \rightarrow Ar_2(^3\Sigma_g^+) + \gamma \rightarrow 2\,Ar + \gamma. \tag{1.17}$$

The singlet transition to the ground state $Ar_2(^1\Sigma_g^+)$ of the dimer is fast ($\sim$ns) while the triplet transition to $Ar_2(^3\Sigma_g^+)$ is suppressed and therefore slow ($\sim\mu$s). As a result, the scintillation light in liquid argon has a slow and a fast component. The ratio of the intensities of the two components depends on the ionization density [22] and hence on the type of particle that ionizes the medium. This can be used for particle identification by pulse shape discrimination [35]. The decays of both, triplet and singlet states, lead to a main emission peak at a wavelength of roughly 128 nm with a spectral width of 7 nm

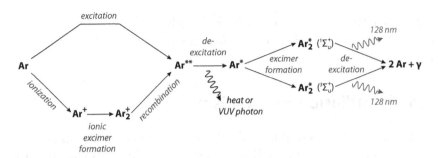

Figure 1.5: Simplified picture of the main scintillation mechanisms in liquefied argon.

to 10 nm [36, 37]. Considered simply, the ground states $Ar_2(^1\Sigma_g^+)$ and $Ar_2(^3\Sigma_g^+)$ are repulsive only and thus very short-lived [22]. They almost immediately separate into two argon atoms (see Eqs. 1.16 and 1.17). As a result, there is no resonant re-absorbtion of the scintillation photons and argon is transparent at these wavelengths.

1.6 Charge carrier transport

In the absence of an electric field, the velocities of free charge carriers in a medium follow a Maxwellian distribution and are isotropic. When an electric field $\vec{E}_d$ is applied, electrically charged particles like ions and electrons are accelerated to reach a certain net velocity, called drift velocity $\vec{v}_d(t) = \langle \vec{v}(t) \rangle = $ const. with a direction along the electric field lines towards the cathode (anode) for positively (negatively) charged particles. Other than in the vacuum, the carriers are not accelerated to higher and higher speeds but are subject to collisions with the surrounding medium. Their mean drift velocity is given by the product of the mobility μ with $E_d = |\vec{E}_d|$. μ is a function of the host medium, the type of charge carrier as well as of E_d and of the temperature T. In the absence of magnetic fields, one finds

$$\vec{v}_d = \mu(E_d, T) \cdot \vec{E}_d. \tag{1.18}$$

Over small ranges in (E_d, T) and for low E_d of up to about $200\,\mathrm{V/cm}$, μ can be considered constant in liquid argon [38, 39], resulting in a linear dependence of v_d on E_d in this region of electric field strengths.

An overview on measurements of $v_d(E_d)$ in liquid argon is given in Figure 1.6. The dependence starts to become non-linear already at electric field strengths below $E_d = 0.5\,\mathrm{kV/cm}$. Moreover, one concludes from the figure that μ decreases with increasing temperature and increasing E_d. At typical E_d for large LArTPCs, v_d is of the order of $1\,\mathrm{mm/\mu s}$ to $2\,\mathrm{mm/\mu s}$. A precise knowledge of v_d is indispensable to ensure the precision in the spatial drift coordinate z measurements in a TPC. Compared to the electrons, the argon cations have a zero field mobility which is lower by a factor of $\sim 10^6$ (see Tab. 1.1), resulting in drift speeds of the order of mm/s at the given field strengths. For large LArTPCs and high event rates, this may lead to the accumulation of positive space charge of increasing density towards the cathode, resulting in distortions of the electric field. This effect is especially of concern for detectors that are located at shallow depths or on the earth's surface as they are exposed to the high flux of cosmic radiation. It must be considered and corrected for to recover the spatial resolution of the device.

Another process affecting the spatial resolution of a TPC is diffusion, which describes the broadening of the spatial distribution of a charge cloud with time. In the case of a zero electric field and for thermalized electrons, diffusion is isotropic and is described by a diffusion constant D. The Einstein equation relates D with the electron zero field mobility μ_0

$$\frac{eD}{\mu_0} = kT, \tag{1.19}$$

where e is the elementary charge, k is Boltzmann's constant and T denotes the temperature. In the presence of an electric field, the mobility is no more given by μ_0 and the electron energy differs from the thermal energy kT. Hence, Equation 1.19 must be modified to [20]

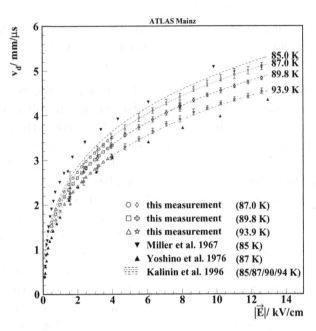

Figure 1.6: A collection of experimental results on the electron drift speed versus the electric field strength in liquid argon at different temperatures [39]. It includes measurements from references [38–41].

$$\frac{eD}{\mu} = f \cdot \langle E \rangle , \qquad (1.20)$$

where $\langle E \rangle$ denotes the mean electron energy and f is a function that depends on the electron velocity distribution. The product $f \cdot \langle E \rangle$ is referred to as the mean electron agitation energy. Apart from leading to a modified Einstein equation, the diffusion process also becomes anisotropic and dominant in the transverse direction with respect to $\vec{E_d}$. Hence, it must be described by two different coefficients D_L and D_T for the longitudinal and transverse directions respectively. Given the diffusion coefficients, the spread of an initially infinitely narrow charge cloud reaches a finite width of

$$\sigma_{L,T} = \sqrt{2tD_{L,T}} \qquad (1.21)$$

after a time t. Drift times in long drift LArTPCs, reaching values of several ms, make diffusion non-negligible limiting the spatial resolution of the device.

1.7 Electron attachment losses

The removal of impurities from noble gases and liquids has already been studied extensively [6, 7, 42–46]. The main contaminants of concern in these media are atoms or molecules with a high electronegativity as they have a larger tendency to attach an electron to form a negative ion. Typical examples of contaminants of that kind present in LArTPCs are oxygen O_2, water H_2O or carbon dioxide CO_2. The attachment of electrons to atoms and molecules happens in various ways [19]. One possibility is radiative attachment, where a neutral atom or molecule binds an electron to form a negative ion under the emission of a photon. Due to its low cross-section, this process is negligible in liquid argon [19]. A second way how quasifree electrons can be temporarily bound is dissociative attachment. A molecule XY captures an electron to dissociate into X and Y^-, either via an intermediate excited XY^* or a negatively ionized XY^- state. In liquefied argon, however, the dominant branch for electron attachment is a three-body process which was first described by Bloch and Bradbury [47] and later refined by Herzenberg [48]. According to this mechanism, electron attachment to a neutral atom or molecule XY proceeds in two stages. Firstly, the XY captures an electron to produce a vibrationally excited temporary negative-ion state $(XY^-)^*$

$$e^- + XY \longleftrightarrow (XY^-)^*. \qquad (1.22)$$

After a finite lifetime, the $(XY^-)^*$ either autoionizes back into a free electron and the neutral molecule or it collides with a third partner, usually an atom or molecule of the host medium and thereby loses its excess vibrational energy

$$\left(XY^{-}\right)^{*} + \text{Ar} \quad \longrightarrow \quad XY^{-} + \text{Ar}. \tag{1.23}$$

The rate of electron removal from a charge cloud consisting of a number $N_e(t)$ of electrons at time t is appropriately modelled by [19]

$$\frac{dN_e}{dt} = -k_{\text{tot}} \cdot N_e(t), \tag{1.24}$$

where $k_{\text{tot}} = \sum_i k_i$ with the k_j describing the probability for an electron to be attached to an atomic or molecular impurity of type j. The k_j are directly proportional to the number densities n_j of the corresponding kind of impurities. By integrating Equation 1.24, one finds that the number of ionization electrons remaining after a time t is given by

$$N_e(t) = N_0 \cdot e^{-k_{\text{tot}}t} = N_0 \cdot e^{-t/\tau}, \tag{1.25}$$

where N_0 denotes the initial number of ionization electrons left after recombination, i.e. at $t = 0$ here. A rule of thumb relating the level of oxygen equivalent impurities to the characteristic time constant τ (often named the 'charge lifetime' in this context) is given by [49]

$$\rho_{O_2} \, [\text{ppt}] \approx \frac{300}{\tau \, [\text{ms}]}, \tag{1.26}$$

where ppt stands for parts per trillion (10^{-12}).

The removal of drifting electrons due to impurities diminishes the charge that can be read out at the sensing plane and thus reduces the signal-to-noise ratio of the detector. Hence, the level of impurities has to be kept as low as possible. Moreover, to obtain calorimetric information, one must have knowledge of the argon purity or, equivalently of τ, to correct for the attenuation of the signals. It is either measured by means of ionizing particles [50–52], from UV laser ionization tracks [6, 7, 53] or from an induced release of charge [52, 54].

1.8 Wireplane read-out and signal formation

A wireplane is an array (grid) of thin conductive parallel wires separated by typically 2 mm to 4 mm from each other. To realize segmentation along different spatial coordinates, the read-out configuration consists of two or three wireplanes of different orientations separated by a small gap of up to a few mm. The left hand side of Figure 1.7 shows the projections y-z and x-z of a read-out configuration with two wireplanes. The lower one with wires oriented along the y axis is labelled X while the upper one is named Y and has wires along the x axis.

An electron generated in the detector volume follows the applied uniform drift field $\vec{E}_0$ and approaches the first wireplane X which is set to ground potential. To make sure that the electrons are not collected already by the first wireplane, the Y grid must be biased to a positive potential to produce an electric field $|\vec{E}_1| > |\vec{E}_0|$ between the two planes. By setting up a field of an appropriate strength, the field lines terminate only on the wires of the Y plane and hence the X plane appears to be fully transparent for the drifting electrons (see Fig. 1.7, left). The transparency condition for a wire grid consisting of equidistant wires and separating two regions i and j with electric field strengths of E_i and E_j respectively is given by [55]

$$|\vec{E}_i| > \frac{1+\rho}{1-\rho} |\vec{E}_j|, \qquad (1.27)$$

where $\rho = 2\pi r/d$ with r and d the wire radius and the wire spacing respectively.

The moving charge carriers approaching the wireplane X induce currents in the wires located at the corresponding spatial coordinates. Once the electrons have passed through the wireplane, the electric current in the wires is reversed. That is why a bipolar signal is observed from the wires of the X plane (see Fig. 1.7, right). As long as the transparency condition is met, none of the drifting charge carriers are collected at the wireplane X and the read-out is said to be

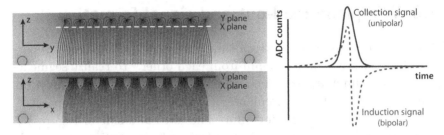

Figure 1.7: *Left:* Typical wireplane read-out of a LArTPC. The X plane is at ground potential and in the non-destructive induction mode, i.e. fully transparent for the drifting electrons. This is achieved by setting the Y grid to an appropriate positive potential. The electric field lines terminate on the Y plane which is thus in the destructive collection mode. *Right:* Examples of bipolar induction (dashed line) and unipolar collection (solid line) signals as they result from the X and Y planes respectively.

non-destructive or operated in induction mode. At the Y wireplane the electrons are finally collected (collection mode, destructive read-out) which is why the induced signals from the wires of the Y plane are unipolar. To resolve ambiguities in the event reconstruction, a setup with three wire grids U, V, X (two in induction and one in collection mode) is usually chosen. The two induction planes U and V are at specific angles with respect to the collection plane. Sometimes, the U plane is not instrumented and only serves as a shielding grid to avoid signal induction on the V plane before the electrons enter the read-out zone. Such a configuration ensures to obtain a well-shaped signal from the V plane.

For a wireplane configuration typically set up for LArTPCs, a track induced by a minimum ionizing particle leads to a number of about 6000 electrons per wire at the read-out [56]. This corresponds to about 1 fC of charge. To process and digitize such signals with an ADC (analog-to-digital converter), an intermediate electronic preamplifier stage is necessary.

2 The ARGONTUBE detector

The next generation of LArTPCs with sensitive masses of the order of 10 kt to 100 kt introduce a number of new challenges. A major one is the long drift distance of 10 m to 20 m needed to reduce the number of read-out channels, and thus the cost, and allowing to build a detector with a large monolithic sensitive volume. An electric field strength of about 0.5 kV/cm to 1 kV/cm was found to be a good compromise to limit the amount of electron-ion recombination, to obtain a high drift velocity and to keep diffusion at a low level. On distance scales of 10 m to 20 m, this results in voltages of up to 2 MV between anode and cathode. It is a difficult task to construct a leak-tight feedthrough that holds these voltages without dielectric breakdowns. An alternative way is to generate the voltage directly inside the cryostat [6–8]. A consequence of high voltages are strong electric fields between the TPC field cage and the cryostat walls. The dielectric strength of liquid argon was measured at distance scales of 100 μm and less, and voltages of the order of 10 kV [57, 58]. Depending on the argon purity, breakdown field strengths reaching from 0.85 MV/cm to 1.7 MV/cm were measured. Reports from currently running LArTPCs and dedicated measurements indicate that these limits are not valid at larger distance scales of several mm or more [59–61].

Apart from the difficulties related to high voltages, larger drift distances imply longer drift times and the level of impurities in the medium becomes more relevant. The attenuation of the detector output signals by contaminant molecules grows exponentially with the drift time. Control and continuous removal of impurities coming from outgassing of detector parts and from residual gas leaks is necessary. By using the latest cryogenic preamplifier technologies, one can partly compensate the signal attenuation and significantly improve the signal-to-noise ratio [56, 62–64].

To study the feasibility of long drift distances of the order of several meters along with the necessary technologies, the ARGONTUBE project has been initiated at the AEC-LHEP of the University of Bern. It is a LArTPC with a drift distance of nearly 5 m, the longest reached to date [6–8]. Apart from the cryostat vessels the detector was designed and constructed in Bern and firstly operated in 2011. It is installed in the 'Grosslabor' in a 7 m deep pit to allow for convenient maintenance and upgrading of the device (see Figure 2.1, left). This chapter gives an overview on the ARGONTUBE cryostat design, the TPC, high voltage generation, liquid argon purification methods, the installed UV laser system, read-out, data-acquisition (DAQ) and triggering systems as well as the routine operation of the detector. It closes with a gallery of events recorded during the latest ARGONTUBE runs.

2.1 Cryostat design

The ARGONTUBE cryostat is realized as a bath cryostat, meaning that it is composed of two stainless steel cylinders, an inner (main) and an outer vessel. The latter is permanently installed in the 7 m deep pit (see Fig. 2.1, left), has a diameter of about 80 cm and a length of roughly 6 m. During detector operation, it serves as a cold reservoir for the inner vessel and is continuously refilled with liquid argon. To minimize the heat input, it features a vacuum insulated wall containing 50 layers of superinsulation film. The inner vessel has a diameter of 50 cm and a length of 5.6 m and contains the TPC. Both, TPC and inner vessel can be completely extracted from their containers for maintenance and upgrades of the detector. During operation, the inner vessel is hermetically sealed. The feedthroughs needed for instrumentation of the device are located on the top flange and are, wherever possible, made with conflat (CF) flanges and copper gaskets which provide a metal-to-metal seal suited for cryogenic temperatures. Top and bottom flanges are sealed by means of indium wire. The outer volume is not hermetically sealed as there is no exchange with the inner volume and thus no need for highly pure liquid argon. The

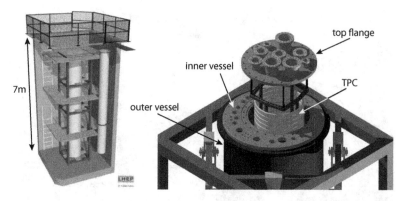

Figure 2.1: *Left:* CAD model of the 7 m deep pit where the detector is installed. *Right:* CAD drawing of the bath cryostat with the outer and the inner vessels as well as the partly extracted TPC field cage attached to the top flange [8].

right hand side of Figure 2.1 shows a drawing of the outer vessel, the inner vessel and the partly extracted TPC field cage attached to the top flange. Two photographs of the real device, during the maintenance phase (left) and during operation (right), are given in Figure 2.2.

2.2 Field cage design and high voltage generation

The TPC field cage has a cylindric geometry of diameter 40 cm and of a length of up to 4.96 m. It is an array of 125 field-shaping ring electrodes (see Fig. 2.3). The simulation software COMSOL was used to optimize the field cage geometry for field uniformity, active detector volume and mass [7]. At the same time, the electric field strength between the cryostat wall and the field cage was minimized to reduce the risk of dielectric breakdowns. Ring electrodes with a racetrack-shaped cross section arranged with a pitch of 4 cm and a gap of 5 mm were found to fulfill the requirements. To achieve a smooth and inert surface, they are made of gold-plated and polished solid aluminum. Given that the electrodes are at their design electric potentials, the

Figure 2.2: *Left:* The ARGONTUBE during the maintenance and preparation phase for the following run. The TPC is attached to the
top flange and partly extracted from the inner vessel. *Right:*
The ARGONTUBE fully instrumented and in operation.

simulation indicates field strength variations smaller than 0.5 % within
the detector sensitive volume [7]. Torlon PAI (polyamide-imide)[1]
pieces are used to combine five rings to one module and to assemble
the field cage as illustrated in Figure 2.3 on the bottom right. PAI is a
high-tech polymer with exceptional mechanical and thermal properties,
chemical resistance and a low outgassing rate. The field cage structure
which weighs nearly 250 kg is attached to the top flange of the inner
vessel by four PAI pillars (see Fig. 2.2, left).

ARGONTUBE was designed for an electric field strength of 1 kV/cm.
Given the drift distance, this requires a potential difference between
cathode and anode of nearly −500 kV. The construction of a leak-tight
feedthrough suited for such a high voltage (HV) is a very challenging
task and a different approach was chosen. Instead of generating
the voltage outside, a Greinacher/Cockcroft-Walton HV multiplier,

---

[1]www.solvayplastics.com

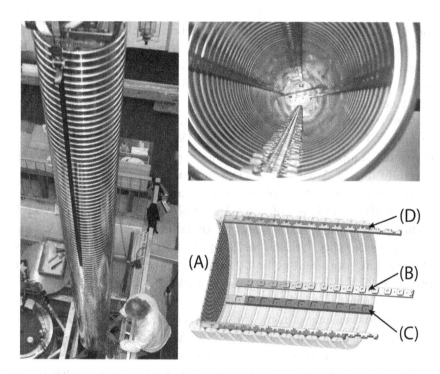

Figure 2.3: *Left:* The fully assembled field cage with 125 ring electrodes. *Top right:* View on the inside of the field cage. Four PAI columns hold the structure together. The Greinacher voltage multiplier is installed on one of the columns. *Bottom right:* CAD drawing showing a cutaway view of the lowermost part of the field cage and the cathode (A). The principle of the assembly is illustrated. PAI pieces (C) hold together the five rings of one module. Their counterparts (B) are plugged into them across two consecutive modules (D).

immersed in liquid argon, produces the HV directly in place. This allows the use of low supply input voltages of the order of a few kV for which commercial feedthroughs are available. A diagram of the first two stages of the Greinacher circuit used for the ARGONTUBE is shown in Figure 2.4. It consists of diodes, capacitors and resistors. The latter are added to limit the discharge current to protect the diodes and capacitors in case of a dielectric breakdown. The circuit is driven by an AC input voltage (VAC) with an amplitude U_{in}^{AC}. The theoretical DC voltage at the final stage, given a number N of stages, amounts to $U_N^{DC} = -2NU_{in}^{AC}$ with respect to the reference point denoted by *Ref* in the figure. In theory, an arbitrarily high voltage can be generated by adding more and more stages to the cascade. To operate the Greinacher HV multiplier successfully at cryogenic temperatures, the electronic components must be chosen carefully [8]. In ARGONTUBE, the current configuration uses $4.7\,k\Omega$ film type resistors with a tolerance of 5 % and a rating of 0.5 W, diodes of type M160UFG manufactured by Voltage Multipliers Inc[2] with a threshold voltage of 15 V and a breakdown voltage of 16 kV, and capacitors of type PHE450 made by Evox Rifa[3] with a capacitance of 47 nF (at room and at liquid argon temperatures) and a voltage rating of 2 kV. The originally used capacitors with a rating of 4 kV and a capacitance of 158 nF (at liquid argon temperature) had to be replaced for reasons of robustness at cryogenic temperatures. With the new type, only half ($-250\,kV$) of the ARGONTUBE design potential can be reached in theory. The actual limiting factor, however, are dielectric breakdowns in liquid argon occurring already much below this value.

Thanks to the cascade structure of the HV multiplier, it was possible to design the field cage such that there is exactly one Greinacher stage in between two ring electrodes. Given the 125 electrodes, a bias voltage of 2 kV between any two rings leads to a maximum negative potential of $-250\,kV$ on the cathode. Consequently, an AC voltage of up to 2 kV peak-to-peak has to be fed in. The Greinacher multiplier with its sharp edges and tiny structures is installed on the inside of the field cage (see Fig. 2.3, top right) to avoid high electric field strengths between the circuit and the cryostat wall and thus to reduce the risk of dielectric breakdowns. The drawback are local distortions of the electric field in the detector sensitive volume.

---

[2]www.voltagemultipliers.com
[3]www.kemet.com

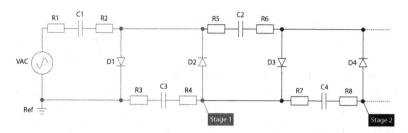

Figure 2.4: The first two stages of the Greinacher voltage multiplier used for the ARGONTUBE to generate the HV. It produces negative DC voltages with respect to the reference point *Ref*.

Compared to using an external power supply to generate the HV, the Greinacher multiplier has the disadvantage of taking a relatively large amount of time to be charged. The characteristic charging time is dependent on the capacitance, the protective resistors and on the charging frequency. Due to a continuous leakage current or occasional dielectric breakdowns in ARGONTUBE, the Greinacher multiplier must be recharged periodically during detector operation. The data taking is vetoed during these periods as there would be a strong pickup noise induced on the sensing wires of the read-out plane. To increase the Greinacher charging speed by a factor of four, the frequency of the AC input has recently been changed from 50 Hz to 500 Hz. At 500 Hz, the characteristic charging time of the circuit was measured to be $\tau = 78 \pm 4$ s.

2.3 Liquid argon purification

For drift distances as long as those in ARGONTUBE, an electron 'charge lifetime' of several ms must be achieved. Following the rule of thumb given in Equation 1.26, this translates into $\rho_{O_2} \approx 0.1$ ppb of oxygen equivalent impurities in the sensitive detector volume. To reach these values, it is necessary to correctly design the system by using standard and well-established vacuum technique [65]. This includes a careful choice of materials[4] with thorough cleaning and baking out before installation to minimize the outgassing rate from their surfaces. Cryogenic temperatures, mechanical stability, dielectric strength and other required properties limit the spectrum of materials that can be used. In terms of vacuum suitability, metals like stainless steel or aluminum are good choices. When metals may not be used for certain reasons, high-tech plastics like PAI, PEEK (polyether ether katone) or materials from the Teflon family should be used. Any material with a high affinity to water, such as nylon, should be avoided. Apart from the material choice, the structures which are located inside the vacuum chamber must be designed such that they do not have any enclosed volumes (virtual leaks) which continuously release contained air at small rates. Moreover, the external leak rate must be held as low as possible and careful leak tests with a sensitive device (helium leak detector) are indispensable. Despite the careful choice of materials, there always is residual outgassing. At cryogenic temperatures, the process is slowed down and the rate of outgassing is considered negligible for detector parts situated in liquid argon. In the argon gas phase at the top of the detector sensitive volume, however, significant outgassing takes place and constantly supplies the argon with impurities. Besides, tiny external leaks are unavoidable (permeation).

To reach a high level of argon purity in ARGONTUBE, the inner vessel is evacuated by a vacuum pump system for at least one week before filling and operation to remove air and moisture from the volume. Commercially available liquid argon contains only about order of ppm oxygen and water contaminants [66]. To reach sub-ppb levels of impurities, however, additional in situ purification of the detector medium is necessary. The first stage of liquid argon purification is done upon filling of the inner vessel when the argon passes through an oxygen and water filter[5]. The working principle of the purifiers

---

[4] outgassing.nasa.gov

[5] Criotec Impianti S.r.l, Via Francesco Parigi 4 – zona industriale Chind, 10034 Chivasso (To), Italy.

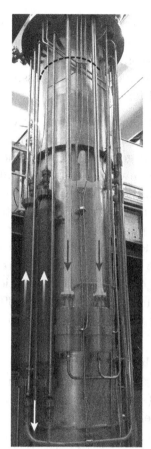
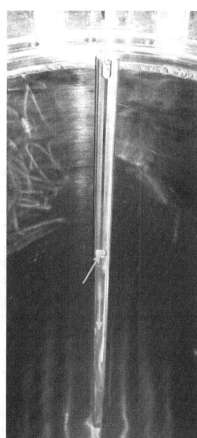

Figure 2.5: *Left:* Liquid argon is taken in from the top of the vessel by two bellow pumps (purple) through the inlet pipe (yellow). It is pushed through the filters (blue) and is brought back to the main volume through the line (red) which reaches to the very bottom of the vessel. The dashed line in the upper part of the image indicates the approximate level of the wireplane. Colors are available online. *Right:* Photograph of the additional pipe that was installed to shift the inlet to the height of the wireplane (cut out window). It is held by a screw on top and stuck into the original inlet at the lower end.

is explained in Chapter 3. During detector operation, the liquid is continuously recirculated through two filter cartridges of the same type, since one-time filtering is not enough to reach the necessary purity levels and the permanent contamination by outgassing and residual gas leaks must be compensated for.

The ARGONTUBE recirculation system was manufactured by Criotec Impianti and is shown in Figure 2.5 on the left. The direction of flux is indicated by the arrows. The liquid argon flux is maintained by two bellow pumps, driven by pressurized nitrogen gas. They take in the liquid and push it through the two filter cartridges. The argon is injected back to the main volume through the pipe reaching to the very bottom of the inner vessel. Bellow pumps were chosen for their high throughput, simplicity and purity, low heat input and low noise induction on the detector read-out. The latter was the major reason as the recirculation has to run non-stop during detector operation to guarantee the necessary level of argon purity. A flux of about $300\,l/h$ is achieved corresponding to a full inner volume change every four to six hours.

The pump inlet is situated well below the inner filling level which is indicated by the dashed line shown in the upper part of Figure 2.5 on the left. Hence, an additional pipe (see Fig. 2.5, right) has been installed on the inside of the inner vessel to shift the inlet up to the level of the wireplanes where most of the contaminants are expected to come from (gas phase). Since there is a gap of only about $5\,cm$ between the TPC field cage and the cryostat wall, a pipe with a flat cross section was chosen and care was taken to manufacture the bending at the lower end as smooth as possible to minimize the electric field strength between the field cage and the pipe. At the level of the newly installed pipe, the absolute electric potentials on the field cage rings are relatively small ($< 50\,kV$) and no additional negative impact on the stability of the HV during detector operation was found as a consequence of this modification.

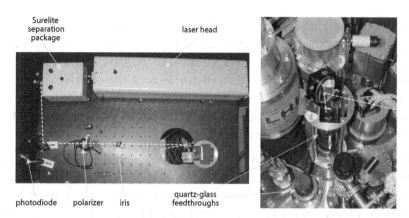

Figure 2.6: *Left:* A photograph of the laser table mounted above the
ARGONTUBE. The approximate laser path is indicated by
the dashed line. *Right:* View on the partly instrumented
top flange with the two quartz-glass feedthroughs and the
mirror used to steer the beam inside the TPC. The light
pulse enters from the right hand side in the figure.

2.4 UV laser system

The ARGONTUBE is equipped with a UV laser system (see Fig. 2.6)
which allows the production of straight ionization tracks in the detector via the process of multi-photon absorption. The light source
is a Continuum Surelite I-10 Nd:YAG [67] pulsed laser[6] which emits
photons of wavelengths 1064 nm along with their harmonics 532 nm,
355 nm and 266 nm. An intense beam of photons at a wavelength of
266 nm is well suited to ionize the argon atoms efficiently by quasi-
resonant ionization [23–25]. By means of the Surelite Separation
Package (SSP) the 266 nm component (fourth harmonic) is selected
and the remaining wavelengths are filtered out. The laser provides
an energy of 4 mJ per pulse of 4 ns to 6 ns length at a maximum
frequency of 10 Hz. The energy stability from shot to shot is specified
to be ±7 % (RMS 2.3 %). The power drift is indicated to be ±3 %

---

[6]Continuum, 3150 Central Expressway, Santa Clara, CA 95051, USA.

over a time period of 8 h at stable temperature. The beam divergence is about 0.6 mrad. Since argon cations have a rather low mobility (see Tab. 1.1), there is an accumulation of positive space charge in the detector volume that causes distortions of the electric field and hence of the observed tracks when running the laser at the maximum frequency. Thus, a pulse rate of less than 0.5 Hz is used for the AR-GONTUBE UV laser measurements.

The laser head is installed in the box shown in Figure 2.6 on the left and is situated above the ARGONTUBE detector. After passing through the wavelength separation unit, the light pulse passes by a sensitive photodiode (Thorlabs DET10A/M), goes through an adjustable polarizer (Altechna Watt Pilot), an iris, and a system of mirrors before it enters the detector through an evacuated quartz-glass feedthrough [24] installed in the center of the cryostat top flange (see Fig. 2.6, right). The photosensor triggers the data acquisition (DAQ) when recording laser induced ionization tracks. The stray light at the second mirror (see Fig. 2.6, left) is intense enough for the photodiode to generate a signal. The polarizer allows adjustment and attenuation of the beam pulse energy to control the strength of ionization in the ARGONTUBE. By tuning the mirrors installed along the laser beam path, the direction and position of the ionization track within the TPC can be controlled. However, the geometry of the quartz-glass feedthrough and the design of the top flange allow only slight variations in position and angle of the beam and it is not possible to systematically scan the entire sensitive detector volume. The quartz-glass feedthrough reaches into the liquid argon by a few cm to avoid unwanted and non-uniform diffraction and reflection effects at the gas-liquid interface in the detector. It is made leak-tight to better than 10^{-8} mbar l/s (std. He). The glass tube is evacuated to avoid formation of mist on the glass surface caused by the low temperatures inside the cryostat vessel.

2.5 Read-out, data-acquisition and triggering systems

The ARGONTUBE setup uses a read-out configuration with two wire-planes oriented at an angle of 90° with respect to each other. The photograph in Figure 2.7 on the left gives an impression of the read-out section of the detector. The plane separating the drift volume from the read-out area is in induction, while the second one situated behind is operated in collection mode. The two planes are equipped with 64 sensing wires each, made of a Cu-Be alloy and separated by 3 mm from one another. They have a diameter of 125 μm and a length of 20 cm. As a result, the sensitive volume of the ARGONTUBE is $0.2 \cdot 0.2 \cdot 4.96 \approx 0.2 \, m^3$, equivalent to a liquid argon mass of 280 kg. To fulfill the transparency condition (see Eq. 1.27) for the induction plane, the wires of the collection plane are biased to a positive DC potential. For that purpose, decoupling capacitors are installed at every wire. The signals coming from the 128 wires are fed to preamplifiers before being digitized by ADCs.

The ARGONTUBE read-out has recently been upgraded from 'warm' preamplifiers [24], which were installed outside the cryostat, to cryogenic ones (LARASIC4 [56, 62, 63]). The latter read-out scheme offers many advantages reaching from easier cryostat design (multiplexing), an important aspect for future large LArTPCs, to a substantially higher signal-to-noise ratio. The noise level is much reduced thanks to the cryogenic temperatures (thermal noise) and less pickup and cable capacitance noise, as there is no long cable connection between the signal source (sensing wires) and the amplifier stage. The LARASIC4 ICs are based on CMOS technology and have been developed particularly for the use in LArTPCs by electronics engineers and physicists at Brookhaven National Laboratory (BNL) [56, 62, 63]. They are located inside the cryostat, immersed in liquid argon and are directly connected to the wireplane. One chip includes both a charge amplifier and a shaper for each of its 16 channels. Amongst others, it has four programmable gain configurations 4.7, 7.8, 14 and 25 mV/fC (charge-

internal PMTs   quartz-glass feedthrough

wireplanes   Greinacher circuit   preamplifiers

Figure 2.7: *Left:* Read-out section of the ARGONTUBE showing the
wireplane with the cryogenic preamplifiers (LARASIC4
chips [56, 62, 63]). The capacitors of the Greinacher circuit
as well as the central quartz-glass feedthrough are visible.
The PMTs are inside a Faraday cage for reasons discussed
in the text. *Right:* A photograph of the two internal PMTs
shielded by an aluminum cage after a test in liquid argon.

sensitive amplifier) and four different peaking time T_p settings 0.5, 1.0,
2.0 and 3.0 μs. The ARGONTUBE read-out is configured for maximum
gain and $T_p = 3\,\mu$s. The transimpedance (current-sensitive amplifier)
for this configuration was measured to be $Z_{\mathrm{cryo}} = 117 \pm 3\,\mathrm{mV/nA}$,
compared to $Z_{\mathrm{warm}} \approx 13.0\,\mathrm{mV/nA}$ for the former 'warm' preamplifiers [68]. Both values originate from testbench measurements. Every
channel features an injection capacitor that can be used for calibration
and tests. A printed circuit board (PCB) was designed at LHEP to
specifically match the requirements of the ARGONTUBE wireplane
configuration. One PCB hosts two LARASIC4. A NIM module supplies power and is used to configure the gain and peaking time of the
chips. It features a test pulse generator such that each channel of the
amplifiers can be checked individually.

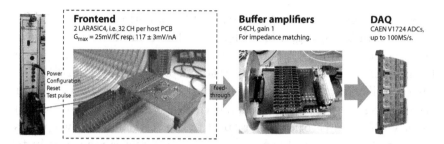

Figure 2.8: Signal chain for the cryogenic read-out. The first device on the left is the NIM module used to power, test, configure and reset the LARASIC4. Starting from the wireplane, the signals are amplified by the LARASIC4, fed out of the cryostat (flat cables), pass through buffer amplifiers (for impedance matching) and are digitized by the ADCs. The dashed rectangle represents the cryostat.

Figure 2.8 illustrates the read-out chain of ARGONTUBE with cryogenic electronics. Starting from the wireplane, the signal pulses are amplified, fed out of the cryostat via flat cables, pass through a buffer amplifier (gain 1, for impedance matching purposes) and are digitized by the CAEN[7] V1724 ADCs with an amplitude resolution of 14 bit and a sampling period of $1.01\,\mu s$. These devices read out and digitize the signals continuously and store them temporarily in a ring buffer. Upon a trigger event, within an acquisition window of $8274\,\mu s$ including a pre-trigger window of $595\,\mu s$, the digitized signals are read out of the ring buffer and sent via optical link to a computer where they are displayed and stored in raw format. For the analysis and reconstruction of the events, the rawdata are converted to ROOT[8] files.

Depending on whether UV laser induced tracks or cosmic events are recorded, the DAQ is triggered respectively by the photodiode (see Fig. 2.6) or by a set of scintillator planes, placed around the AR-

---

[7]www.caen.it
[8]root.cern.ch

GONTUBE, combined with two internal photomultiplier tubes (PMTs) immersed in liquid argon (see Fig. 2.7). The scintillator planes have an area in the range of about $0.2\,\mathrm{m}^2$ to $0.4\,\mathrm{m}^2$ each and are given by a plastic scintillator (NE102), read out by Philips XP2020 PMTs. Five scintillator planes are distributed around the ARGONTUBE outer cryostat vessel to guarantee angular coverage for cosmic muons with slightly tilted trajectories with respect to the vertical axis. Two planes are installed directly above and below the detector respectively to trigger on passing-through particles. Finally, another two scintillator planes are placed on the side of the detector, slightly apart from the vessel to record cosmic particles with highly tilted trajectories. The two internal PMTs are two-inch Hamamatsu[9] R7725-mod with a maximum response at a wavelength of about 420 nm [69], where the quantum efficiency is about 20 % [70]. They are well suited for the use in cryogenic environments [70]. In ARGONTUBE, they are operated at a supply voltage of 1300 V to 1500 V (max. rating 2000 V) and are placed directly above the wireplane as visible in Figure 2.7 on the left. During detector operation, their lower part is immersed in liquid argon while the upper part, composed of the voltage divider, is in the gas phase. The maximum emission peak of the scintillation light produced in liquid argon is at a wavelength of 128 nm. To match the response of the PMTs and to increase the light collection efficiency, their windows are coated with tetraphenyl butadiene (TPB) – an organic compound that acts as a wavelength-shifter. It absorbs light in a broad range in the deep UV and has its maximum emission peak at $425 \pm 50\,\mathrm{nm}$ [71]. Due to aging effects [72], the TPB coating has to be renewed periodically. As illustrated by the photographs in Figure 2.7, the PMTs are contained in aluminum cylinders. A fine copper mesh with a high transparency is placed at the front cap of the cylinders to let the light pass through. The PMTs are fully shielded to avoid a strong pickup signal on the sensing wires induced by the current pulses within the tubes triggered by the scintillation light. The voltage divider of the PMTs had to be optimized to work stably at the

---

[9]www.hamamatsu.com

desired voltage inside the aluminum containers. During ARGONTUBE operation, the PMT divider is located in the argon gas phase. The dielectric strength of gaseous argon is about twice lower than that of air at the given length scales and conditions [73]. This made extensive tests necessary to prove the PMTs to work at the necessary voltage without the divider breaking through the argon gas to the wall of the shielding cylinder. The tests were done in a small dewar containing liquid argon to simulate the situation in ARGONTUBE.

2.6 Detector operation

The ARGONTUBE is a research and development project and is constantly subject to modifications. Thanks to the modular design and extractability of the TPC, upgrades can be done quickly. To carry out modifications, however, detector operation has to be stopped and all the liquid argon has to be removed from the detector volumes – a process that is accelerated by the use of resisitive heaters installed on the bottom flange of the inner vessel. After completion of the detector upgrades, cleaning and baking out is done and all the instruments are tested. Finally, the inner volume is closed by sealing the top flange with indium wire.

Before starting an ARGONTUBE run, the inner volume is evacuated for at least one week to remove air and moisture. This is done by a permanently installed high-throughput vacuum pump system consisting of a Roots pump (Alcatel RSV 601B) in series with a rotary vane pre-vacuum pump (Alcatel 2063H). By design, this pump system evacuates down to about 10^{-4} mbar [74]. Within one week and given that the system is leak-tight, the inner ARGONTUBE vessel is evacuated to a residual pressure of about $5 \cdot 10^{-4}$ mbar. Apart from removing moisture and air, one can conveniently check the system for external leaks at this stage. Even though the apparatus is operated at a slight argon overpressure and at cryogenic temperatures during the runs, vacuum tightness at room temperature already rules out a large number of possible gas leaks. The gas leaks can

be tracked down by means of a helium leak detector (Œrlikon Leybold Vacuum PHOENIXL 300 [75]) which is sensitive to leakage
rates as little as $5 \cdot 10^{-12}$ mbar l/s (std. He). Usually, values between
10^{-9} mbar l/s (std. He.) to 10^{-8} mbar l/s (std. He.) are achieved for
the inner ARGONTUBE vessel.

To put the detector into operation, the inner and outer vessels have
to be filled with liquid argon. Nearly 2700 l are needed in total, that
is including the amount which evaporates during the cool down of the
entire structure. Filling starts for the outer vessel directly from the
4500 l liquid argon reservoir installed on the outside of the building.
The liquid is transported by vacuum insulated lines. The argon gas
that is generated as a result of boiling, caused by the constant heat
input and the detector structure being initially at room temperature,
is released to the atmosphere through a pipe that goes outside the
building. During this period, the inner vessel is still being evacuated
and as a result of the cool down of its walls, a drop in the pressure
is observed. After about 3 h and decoupling of the vacuum pumps,
the filling of the inner volume starts and the liquid is passed through
the oxygen and water trap as a first stage of purification. The initial
phase of filling the inner vessel is important. The exhaust is given by a
unidirectional valve of relatively small dimensions, enough for normal
detector operation, but too small during the filling procedure as there
are large amounts of argon gas produced by boiling. The additional
bypass exhaust valve is opened only as soon as a high gas flux is
established and an overpressure (≈ 150 mbar) is reached to avoid a
backflow of air into the system. At the beginning, it is also necessary
to adjust the rate of filling. As soon as stable conditions are reached,
the filling valve can be fully opened. In the case of an overpressure of
more than about 300 mbar in the inner vessel, a pressure relief valve
opens for reasons of safety. The whole filling procedure takes a total
time of about 9 h to 10 h. It is continuously monitored by pressure
gauges, capacitance-based liquid level meters and PT100 temperature
sensors located in the inner and in the outer vessels. Additionally,
a gas flow meter is installed at the exhaust of the inner vessel. At

the end of the procedure, the bypass exhaust valve is closed and the refilling of the outer bath is set to the automatic mode. As soon as the level of the liquid argon in the inner vessel stabilizes, its exhaust is closed and the volume is fully sealed.

A detector run usually endures one to two weeks. During this time, the modified components and systems are tested by recording UV laser induced tracks and cosmic events (see Section 2.7). Due to a continuous leakage current of the Greinacher multiplier circuit, the electric potentials on the electrode rings of the field cage decrease continuously and the electric field strength changes over time. Moreover, there are occasionally dielectric breakdowns through liquid argon, presumably from the TPC field cage and the cathode to the cryostat vessel, and the circuit gets strongly discharged. To allow efficient acquisition of cosmic events and to guarantee that they are recorded at the same electric field strength, an automatic mode for the recharging of the Greinacher circuit was realized in LabVIEW[10]. Every few minutes, the Greinacher charging current is measured for several seconds. If the integrated amplitude of the current third and fifth harmonics exceeds a certain threshold, the circuit is considered discharged and it is recharged until the current goes below the given threshold. Due to a strong pickup noise on the sensing wires, the data acquisition is automatically vetoed during recharging periods.

2.7 Event gallery

The following images show a selection of events that were recorded with the ARGONTUBE during the latest runs in June and October 2013. They were taken with the TPC shortened by 20 cm from 4.96 m to 4.76 m and are neither corrected for attachment nor for recombination losses. Two axes are given along the abscissa. At the bottom is given the spatial drift coordinate z and on top the corresponding drift time t_d. The relationship between the two scales is non-linear due to electric field disuniformities in ARGONTUBE. It has been found by

---

[10]switzerland.ni.com

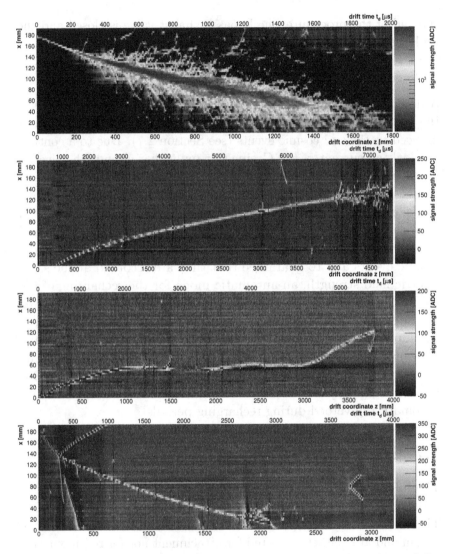

Figure 2.9: *Top to bottom:* Collection wireplane views of an almost fully
contained electromagnetic shower, a full-length (4.76 m) cosmic muon track, a muon stopping in the detector sensitive
volume and decaying to an electron, and a hadronic event.

the methods discussed in Chapter 4. The events are only corrected for disuniformities in the longitudinal electric field component. The residual curvature is attributed to the parasitic transverse field and particle scattering in liquid argon. Note that the aspect ratios of the figures strongly differ from that of the real detector.

3 Regeneration system for argon purifiers

During the preparation phase of the ARGONTUBE run taking place in April 2013, gas leaks were discovered at the bonnet CF flanges of the outlet valves of the two hydro- and oxisorber purifier cartridges (see Fig. 3.1) that are installed in the ARGONTUBE recirculation system. To refurbish the valves, they had to be removed from the purifiers. For this operation, a strong and continuous argon gas flux was maintained through the cartridges to avoid the contamination and saturation of the filters by the surrounding air. Nevertheless, it was unknown whether the gas leaks had led to a strong accumulation of oxygen and moisture in the purifiers as there is no reasonable way to check their condition. Since the purification system is highly relevant for the performance of the ARGONTUBE, the filters had to be regenerated in any case. Having them regenerated by an external company is time-consuming and cost-intensive and, consequently, they were regenerated in-house. Within two weeks, a regeneration system was realized following descriptions given in [42, 43, 76]. After briefly introducing the working principle of the ARGONTUBE purifiers in Section 3.1, the regeneration system and the procedure are explained in detail in Section 3.2. For illustrative purposes, a number of observations and measurements are reported in Section 3.3.

Figure 3.1: A photograph of one of the argon purifiers that are installed
in the ARGONTUBE recirculation system. Both, filter in-
and outlet are closed by all-metal angle valves[1] which are
compatible with cryogenic liquids.

3.1 Working principle of argon purifiers

The level of impurities present in the sensitive volume of a LArTPC
is a critical parameter, particularly for long drift distances and drift
times. Electronegative atoms and molecules diminish the signal as
they tend to attach electrons and remove them from the drifting
charge cloud. Regarding LArTPCs, the main contaminants that cause
signal attenuation are oxygen and water. To remove them from the
medium, different techniques are used. They are usually based on
adsorption processes, that is by adhesion of the contaminant atoms
or molecules to the surface of an agent material, called adsorbent.
Adsorption may happen via two different mechanisms, chemisorption
or physisorption [77, 78]. Concerning chemisorption, adhesion is
guaranteed by a chemical reaction between impurity and agent which
generates a chemical bond. Physisorption on the other hand, works
through binding via van-der-Waals forces to the surface of the agent.
Other than for chemisorption, the agent is not chemically modified
in this case. The agent material used in gas or liquid purifiers is
preferably in a highly porous form to maximize the surface for highest
contaminant capacity. Examples of physisorption agents commonly
used for commercially available filters are molecular sieves, silica gel
(SiO_2) or activated carbon. A typical chemisorption agent for the

---

[1]MDC Vacuum Products LLC, 23842 Cabot Blvd, Hayward, CA, 94545-1661,
USA.

removal of oxygen from a gas or liquid stream is activated copper, often in sintered form, again for reasons of surface maximization.

The purifiers utilized for the ARGONTUBE contain both activated copper to remove oxygen by chemisorption, and silica gel to adhere water molecules by physisorption. Oxygen removal from a gas or liquid stream passing through the filter occurs via the process of oxidation which is accompanied by the emission of heat (exothermic)

$$2\,Cu + O_2 \rightarrow 2\,CuO + heat. \tag{3.1}$$

The amount of oxygen that can be bound in a purifier cartridge is limited. A standard copper agent (BASF R-3-11) has been measured to have a specific oxygen capacity of $C_{O_2} = 4.72 \pm 0.67$ nom. l/kg, which was found to be about 10 % of the theoretical value [43]. The oxidation process can be reversed and thus the filtering capabilities of the agent can be recovered by flushing the purifiers with hydrogen gas H_2 while providing a certain amount of heat. The endothermic reaction taking place is formally given by

$$CuO + H_2 + heat \rightarrow Cu + H_2O, \tag{3.2}$$

with water vapour being a by-product of the reduction process. Similarly, the water molecules that are adhered to silica gel are released by baking the filter out at about 100 °C for a certain amount of time. Hence, purifiers of that kind are fully regenerable.

3.2 The regeneration system

The concept of the regeneration system was to keep it simple and adaptable. It was built using KF flange connections wherever possible since they allow for extendability of the setup while guaranteeing leak-tightness. References [42, 43, 76] describe in much detail the components that are necessary and were extremely useful during the design and realization of the system. Figure 3.2 shows the final setup schematically. Basically, hydrogen gas and heat have to be provided in

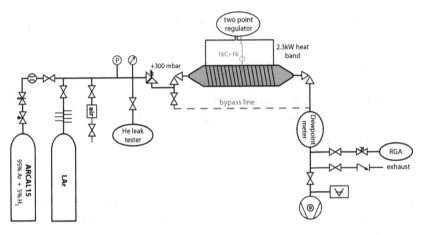

Figure 3.2: Schematic of the regeneration system for argon purifiers.
Different gas supplies are shown on the left. Heat is provided
by a heatband which is controlled by a two-point regulator.
Flux, pressure and moisture content are monitored during
the procedure. The vacuum pump station shown on the
bottom right makes the evacuation of the system possible.

order for the reduction process to take place (see Eq. 3.2). Hydrogen
concentrations of only 4.1 %vol. in dry air at standard conditions mark
the lower limit for flammability and a spark of relatively low energy
may already ignite this kind of gas mixture [79]. Consequently, the
filter is not flushed with pure hydrogen, but with ARCAL15 [80][2],
composed mainly of argon ($95.0 \pm 0.5\,\%$) and only of a small amount
of hydrogen ($5.0 \pm 0.5\,\%$). A pressure reducer and regulation valve is
installed at the gas bottle (200 bar) to adjust a slight overpressure of
about $+100\,$mbar which was found to be enough to ensure the gas flow
through the filter. It has been shown that the regeneration process
is efficient already at a gas flux of about $0.5\,$l/min and there is no
gain in increasing it to larger values [43]. Pressure and flux are monitored by means of appropriate and calibrated devices (Hastings Mass
Flowmeter HFM-200H, Balzers AG TPR 010 and IKR 050 pressure

---

[2]www.carbagas.ch

gauges). A pressure relief valve is installed to open at $+300$ mbar overpressure for reasons of safety. Heat is provided by a glass silk heatband (Isopad IT-S20[3]) which is tightly wrapped around the filter for good thermal contact. Its nominal heating power amounts to 2.3 kW and it is controlled by a two-point regulator (Isopad ICON-TD-7000) that continuously receives a temperature feedback from a temperature gauge (Isopad TAI/NG NiCr-Ni) placed directly on the surface of the filter. Both, filter and heatband are covered with an insulating blanket to reduce the loss of heat. Before starting with the regeneration procedure, the entire system is evacuated through the bypass line (see Fig. 3.2, dashed line) by means of a vacuum pump station (Pfeiffer TSH 260D) including a turbomolecular, and the corresponding backing pump. Gas-tightness of the system is checked by means of the helium leak tester (Œrlikon Leybold Vacuum PHOENIXL 300). During the regeneration phase, monitoring the moisture content at the filter outlet is necessary to check whether the reduction process takes place and to know when it is complete. The first tests were done using a residual gas analyzer or RGA (Extorr XT200[4]) – a mass spectrometer – to monitor the gas composition at the filter outlet over time. RGAs are sensitive devices and must be operated at pressures of the order of 10^{-7} mbar or lower. A leak valve allows fine regulation of the gas flux into the mass spectrometer volume which is continuously evacuated by a turbomolecular pump (not shown in the schematic) to maintain the high vacuum. After a number of first tests and measurements with the RGA, the regeneration system was extended by an additional gas monitoring device, called a dewpoint meter (Shaw Grey Spot sensor and Superdew 3)[5], able to measure the amount of moisture to ppmV precision, and with a sensitivity of 1 ppmV [81]. While the RGA is able to monitor various components of the gas stream simultaneously, the dewpoint meter offers certain advantages when one is interested only in the water content. Reasons are discussed in Section 3.3. After having passed the

---

[3]www.wisag.ch

[4]www.extorr.com

[5]www.shawmeters.com

monitoring devices, the gas is directed through a uni-directional valve to the exhaust where it is released to the atmosphere. Besides the argon-hydrogen mixture used for the reduction process, the system is also connected to the liquid argon reservoir via an evaporator to allow flushing the purifier with pure argon gas [66] at the end of the procedure.

The regeneration procedure starts from an evacuated and leak-tight piping system. The hydrogen-argon gas flux is adjusted to a value between 0.5 l/min to 1 l/min and is guided through the purifier cartridge. As soon as the gas flux and the water content at the filter outlet reach stable conditions (baseline), the heatband is switched on to raise the temperature. Although it may be set directly to the target value lying between 215 and 250 °C [42, 43, 76], it is recommended to increase it slowly (within 30 min) and to wait occasionally for the entire filter cartridge to take on thermal equilibrium. In case the filter is saturated and has accumulated a lot of oxygen, the moisture content at the filter outlet rises significantly already at temperatures as low as 150 °C, indicating the beginning of the reduction process. The runtime and the amount of argon-hydrogen gas needed for the regeneration procedure to finish, depend on the number of contaminant molecules that have been adsorbed by the copper agent. It was found that once the moisture content at the outlet is past the maximum, it falls asymptotically to the initial baseline level. After about 900 cartridge volume changes [42], which takes about 10 h, the gas flux is stopped. There are two possibilities to finish the regeneration procedure and to remove the residual moisture from the filter volume. Either it is flushed with pure argon gas for a few minutes while the cartridge cools down, or, it is evacuated using the turbomolecular pump for 24 h with the heatband switched on [42]. Both methods have been found to work but they have their risks. When flushing with argon gas, it must be very pure as otherwise the filter will again accumulate impurities. When choosing the evacuation method, vacuum tightness is essential to prevent surrounding air from entering the system and spoiling the filter again. To prepare the newly regenerated filter for storage, it is filled with pure argon gas and pressurized to 1.3 bar at the end of the procedure. The slight overpressure prevents the surrounding air from entering the cartridge in case there was a tiny gas leak.

3.3 Measurements and observations

The first tests of the regeneration system were carried out with one of the ARGONTUBE purifier cartridges (see Fig. 3.1) of which the in- and outlet valves had been refurbished beforehand. At that time, the dewpoint meter was not yet available and the RGA was the only device used to monitor the gas stream at the filter outlet. The RGA being a mass spectrometer can only discriminate the molecules against their mass over charge ratio m/z. The ratios that were monitored during the regeneration procedure are shown as functions of time in the upper part of Figure 3.3. They are $m/z = 2$ and $m/z = 18$. They are normalized to the signal strength of $m/z = 40$. $m/z = 2$ and $m/z = 18$ are largely attributed to hydrogen H_2 and water H_2O respectively, while $m/z = 40$ mainly corresponds to ^{40}Ar, which was the dominating component in the gas stream. The idea of normalizing $m/z = 2$ and $m/z = 18$ to $m/z = 40$ was to compensate for any changes in the gas flux from the supply. There was no interest in the absolute pressure values, but only in their relative changes over time which is why the device had not been calibrated. The spikes that are occasionably visible are not caused by a change in the gas composition, but are related to the heatband temperature controller switching an electric current of 10 A affecting the sensitive RGA device. Since a mass spectrometer measures m/z, it is possible that a certain ratio corresponds to more than one molecule type. For instance, the $m/z = 18$ peak is not only caused by water molecules present in the gas stream, but some amount must also be attributed to doubly ionized ^{36}Ar which is one of the argon isotopes. Moreover, $m/z = 2$ does not only originate from pure H_2 but partly results from fragmentation of other molecules like H_2O. The effects of multiple ionization and fragmentation create some ambiguities which can be resolved by thorough analysis, given a

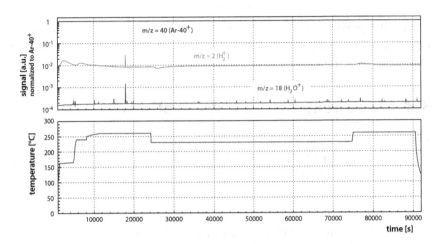

Figure 3.3: *Top:* Time evolution over a period of more than 24 h of the m/z peaks 2 and 18, both normalized to peak 40. No indications of regeneration are visible. Possible reasons are discussed in the text. *Bottom:* Temperature evolution (read out manually) with the sensor located at the filter outer surface.

high enough accuracy in the m/z measurement [82]. The lower part of Figure 3.3 shows the corresponding evolution of the temperature. The periodical variations around the temperature setpoint originating from the operating principle of the two-point regulator are not visible for the reason that the temperature was read out manually. If the regeneration procedure had worked, one would have expected an increase in the water content ($m/z = 18$) as soon as the temperature approached 150 °C. This would have been accompanied by a decrease in the hydrogen content. Due to fragmentation of water molecules, however, it was not a priori clear whether a decrease in H_2 would be visible from the $m/z = 2$ peak. The measurements do not show any additional water vapour at the filter outlet even for temperatures well above those needed to cause efficient reduction of CuO. This may have different reasons. Firstly, the additional water content coming from the reduction process may be buried in the background of the

Figure 3.4: A photograph of the small purifier cartridge originally used for the recirculation system of MEDIUM ARGONTUBE [24].

moisture originating from the RGA or from the pipe walls. Secondly, the filters were far from a saturated state and there was not much copper in oxidized form to be reduced. Or thirdly, the regeneration procedure did not work as expected. Since the purification of liquid argon is essential for the ARGONTUBE to perform well, the procedure and the results shown in Figure 3.3 were further investigated. The idea was to use one of the small purifier cartridges (see Fig. 3.4) of the MEDIUM ARGONTUBE project, a much smaller prototype TPC, constructed at LHEP-AEC [24], for deeper studies. These filters were excellent candidates for tests as they have also been manufactured by Criotec Impianti and are based on the same technology as the large ones. Since they were no more in operation, they could be used to gain more experience with the regeneration system and the procedure.

To compare the filter capabilites for oxygen and water removal before and after the regeneration procedure, a small volume of air (about 10 ml) was pushed once through the bypass line (see Fig. 3.2, dashed line) and once through the cartridge. Using the RGA, the mass spectra of the gas streams at the outlet were measured for the two situations. Zoomed extracts of the spectra are shown in Figure 3.5. The dashed and solid lines correspond to the gas mixture that passed through the bypass line and the filter (before regeneration) respectively. Special care was taken to carry out the measurements under the same conditions. This involved evacuation of the lines and of the RGA volume down to a stable level as well as adjustment of the RGA leak valve to the same leakage rate. Two spectra were measured that looked almost identical. The curves are normalized to the N_2^+ peaks to cancel out the dependence on the amount of air in the RGA volume (total pressure) for the two measurements. The N_2^+ peak was chosen for several reasons. First of all, the purifier does not remove nitrogen molecules from the gas flow and hence when the experiment is done under the same conditions, a peak of the same amplitude is expected independent of whether the air is guided through the bypass line or the filter. Secondly, the $m/z = 28$ peak can be mainly attributed to N_2^+, i.e. there is no large contribution originating from fragmentation or multiple ionization of other molecules present in the system. Last but not least, the nitrogen content in air is large leading to a pronounced peak in the spectrum. The fact that the two spectra look almost identical was due to a saturated purifier. It was saturated already before the measurement as the oxygen content is nearly identical for the two measurements (compare peaks at $m/z = 16$ and 32). The peak at $m/z = 18$, which is mainly attributed to water, is slightly reduced and it is probable that the filter still had some capabilities for water removal. Also, an example of fragmentation is clearly visible from these measurements for the peak at $m/z = 17$. It corresponds to the water molecules of which one of the hydrogen atoms was stripped off, i.e. OH^+ during the ionization process in the mass spectrometer. Similarly, a certain contribution to the signal at $m/z = 16$ is caused by water molecules of which both hydrogen atoms were stripped off. An example of double ionization is the peak at $m/z = 20$, mainly originating from doubly ionized ^{40}Ar.

The fact that the small cartridge was found to be fully saturated represented ideal prerequisites to prove the regeneration procedure to work. The same m/z peaks 2, 18 and 40 were recorded and their evolution over time is shown in Figure 3.6 along with the corresponding filter temperature. This time, a significant drop in the H_2 content is observed as soon as the temperature approaches $150\,°C$ – an indication that the reduction process, which consumes H_2, and thus regeneration, takes place. At the same time, one would expect the amount of

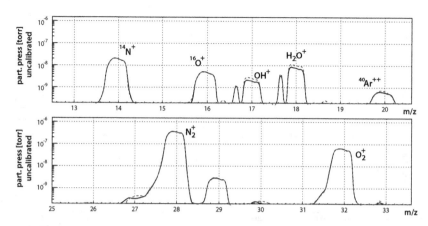

Figure 3.5: Comparison of the m/z spectra for a small air volume having passed respectively through the filter (solid line) or the bypass (dashed line) before purifier regeneration. The two measurements are normalized to the amplitude of the N_2^+ peak. The prominent m/z peaks are labelled with their main contributor.

moisture to increase immediately as soon as the reduction process sets in. However, the data indicate a time delay between the drop in H_2 and the rise in H_2O. The main reason originates from how the RGA measurement was realized. The device was installed relatively far away from the filter outlet (about 1 m) and the gas rate into the RGA volume was strongly limited by the leak valve. Consequently, the main gas flux directly left the system through the exhaust while only a small fraction streamed into the RGA volume. Furthermore, measuring a gas flux which is subject to fast changes in composition with the given RGA is difficult as the evacuation of the former gas composition from the RGA volume takes a significant amount of time. As a result, a fast change in composition only slowly changes the gas mixture inside the RGA volume making the transition appear smooth. About 4 h after the rise, the moisture content starts to decrease asymptotically to almost its initial value. The procedure lasted slightly more than 15 h.

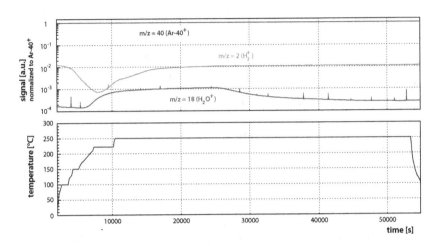

Figure 3.6: *Top:* Time evolution over a period of more than 24 h of the
m/z peaks 2 and 18, both normalized to peak 40. Clear
signs of the reduction process taking place are visible. The
hydrogen content drops as soon as the temperature approaches 150 °C while the amount of water at the filter
outlet increases. *Bottom:* Temperature evolution (read out
manually) with the sensor located at the filter outer surface.

To make sure that the oxygen and water filtering capabilities had
successfully been recovered, the measurements shown in Figure 3.5
were repeated in the same manner as described before. The results are
presented in Figure 3.7. Again, the spectra given by the dashed and
solid lines represent the gas composition after passage through the
bypass line and the (regenerated) purifier respectively. The curves are
again normalized to the amplitude of the N_2^+ peaks. This time, the
oxygen content in the air volume guided through the regenerated filter
is strongly reduced, visible from the $m/z = 16$ and $m/z = 32$ peaks
originating mainly from singly ionized ^{16}O and singly ionized $^{16}O_2$
molecules respectively. The results prove that the filtering capabilities
have been recovered. All the components that are not assumed to
be influenced by the presence of the filter, such as $^{40}Ar^{++}$ (peak
20), $^{40}Ar^+$ and CO_2^+ (not shown) remain indeed unaffected. The

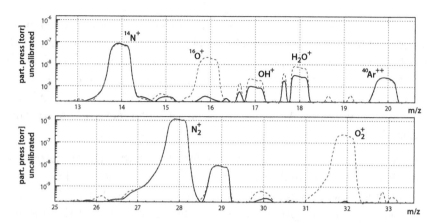

Figure 3.7: Comparison of the m/z spectra for a small air volume having passed respectively through the filter (solid line) or the bypass (dashed line) after purifier regeneration. The measurements are normalized to the amplitude of the N_2^+ peak. The prominent m/z peaks are labelled with their main contributor.

capability for water removal was also slightly improved compared to the measurements made before regeneration (see Fig. 3.5). It could presumably be made even better when evacuating the filter for a long period of time (24 h [42]) while keeping it hot to remove the residual water molecules still present in the filter after the regeneration procedure. For this test, the filter was only flushed with pure argon gas for roughly 10 min at the end of the procedure and the residual water molecules were probably adsorbed back on the silica gel.

Although it is well suited to analyze the composition of a gas mixture, the RGA has several drawbacks as a moisture monitor (fragmentation and multiple ionization) and it makes sense to use a dedicated device instead. References [42, 43] suggest the use of a dewpoint meter which is able to measure the water content in a gas stream to ppmV precision. This device uses a thin aluminum wire which is coated by a hygroscopic layer and an additional porous gold layer on top. The gold layer and aluminum wire form a capacitor. The water molecules enter the porous gold layer, change the dielectric constant and hence cause a change in capacitance[6]. Such a device is easy to use and can be calibrated fast. Due to its working principle, it is not biased by other gas components. In addition, it can be safely operated at atmospheric pressure and hence, the gas flux can be fully guided through its volume. As a result, it has a fast reaction time of the order of 1 s.

Figure 3.8 shows the measurements for the small purifier cartridge recorded during the final tests of the regeneration system. The graph on top shows the measurements obtained with the RGA using the same m/z peaks as for the previous measurements. The plot in the middle illustrates the amount of moisture (in ppmV) as measured with the dewpoint meter. The graph at the bottom shows the corresponding filter temperature. The advantages of the dewpoint meter compared to the RGA are clearly visible when comparing the upper two plots. The RGA indicates a drop in H_2 at a time of about 1500 s which originates from the reduction process becoming more and more efficient at temperatures of 150 °C and higher. While an increase in the water content is not seen from the RGA measurements for a relatively long period of time, the dewpoint meter shows a clear signal of additional water vapour at the outlet starting at about $t = 2000$ s. The moisture content reaches a plateau which is due to the temperature being held constant at 160 °C. As soon as the temperature is raised to the final setpoint of 260 °C, a strong increase is observed in the amount of water vapour. It reaches a maximum at $t \approx 6000$ s and starts to drop asymptotically to the initial value. The oscillations that are visible in the dewpoint meter signal are caused by the temperature fluctuations (not shown in the temperature vs. time plot), originating from the operating principle of the two-point regulator. The temperature varies in a range of roughly 10 °C around the setpoint. This measurement proves the sensitivity and the fast reaction time of the dewpoint meter. For reasons mentioned earlier, the RGA is not able to resolve the fluctuations. As soon as the heaters are switched off (gas flux

---

[6]www.shawmeters.com/sensor.html

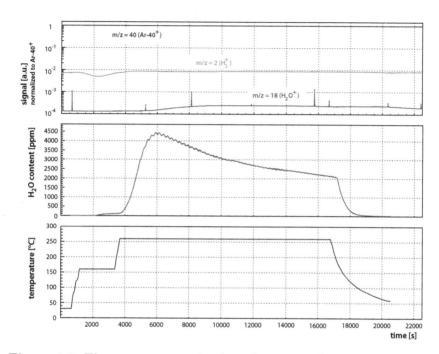

Figure 3.8: The upper two graphs show the time evolution of the gas
composition ($m/z = 2$ and 18 normalized to $m/z = 40$) and
the moisture content at the filter outlet measured with the
RGA (top) and the dewpoint meter (middle) respectively.
The plot at the bottom shows the corresponding temperature, measured at the filter surface and read out manually.

is still on), the moisture content starts dropping to its initial value.
The reason for stopping the procedure at that point was that the
setup should not be left unattended overnight. Obviously, there was
still a relatively large amount of water coming out of the filter and
thus the procedure was restarted the day after. At the end, the
purifier cartridge was evacuated for several hours while keeping it at
a temperature of $260\,°C$ to remove the residual water molecules from
its volume. Finally, it was filled with argon gas and pressurized to
$1.3\,bar$ for storage.

4 The Greinacher high-voltage generator

As explained in Section 2.2, the high voltage needed to operate the ARGONTUBE detector is generated directly inside the cryostat by means of a Greinacher voltage multiplier. Simulations of the ARGON-TUBE field cage indicated a high uniformity of the drift field within the sensitive volume [7]. During detector runs however, it became evident that the electric field was not as uniform as expected from the simulation. The ionization tracks generated with the UV laser are initially straight, but got distorted during the drifting process and arrived at the read-out plane with a strong curvature. The same effect was observed for cosmic muons. Despite the fact that many of the recorded events were induced by high-energy cosmic muons (several GeV), the curvature of their measured ionization tracks was not only much stronger than expected from scattering with the argon atoms, but also systematic. For certain regions of the sensitive volume, the track distortions were more pronounced than for others. It was not a priori clear whether the distortions were due to longitudinal disuniformities of the electric field or due to a parasitic transverse field component. Hence, one is interested in decoupling the two from each other and quantifying the strength of distortion.

Cosmic rays constantly passing through the detector volume cause ionization and generate argon cations at a high rate. Combined with the low ion mobility, this leads to an accumulation of positive space charge resulting in distortions of the electric field. However, from studies of UV laser induced events, and using an appropriate model for the Greinacher circuit (discussed in Section 4.1), one finds that the longitudinal electric field disuniformities can be explained well by the Greinacher multiplier being in an only partially charged state. Partly due to a constant leakage current observed during operation, the Greinacher circuit in ARGONTUBE never reaches the fully charged state. The model proves that in this state, with an increasing number of cascade stages, the generated electric potentials deviate more and more from their setpoints. As a result, the longitudinal (along the drift direction) electric field component is not a constant, but decreases towards the TPC cathode. Moreover, the circuit is installed on the inside of the ARGONTUBE field cage (see Fig. 2.3) and hence it generates (locally) a parasitic transverse component (perpendicular to the drift direction) of the electric field leading to additional distortions of the ionization tracks.

From artificially produced straight UV laser ionization tracks one can extract information about the strength and direction of the electric field disuniformities. By making a full scan of the sensitive volume with the UV laser, the field disuniformities can be measured at every location in the detector and the recorded events can be corrected accordingly. Concerning ARGONTUBE, the currently installed UV laser system does not allow a full volume scan and the electric field cannot be reconstructed in all detail at every position. Nevertheless, the UV laser tracks that have been acquired can be used to decouple the longitudinal from the transverse electric field component, and to find an estimate for the strength of the parasitic transverse field in ARGONTUBE. Combined with the model for the Greinacher circuit, the longitudinal electric field strength can be calculated everywhere in the sensitive volume and thus the drift time can be translated into the drift spatial coordinate. By doing that, cosmic muon and UV laser tracks are corrected for the longitudinal field disuniformities and interesting measurements become accessible, some of which are presented in Chapters 6 and 7.

4.1 Model for the Greinacher circuit

To understand the electric field disuniformities observed in ARGON-TUBE, a model for the charging behaviour of the Greinacher circuit is required. In particular, it is necessary to know the electric potential of each of the field shaping electrodes of the TPC as this allows the calculation of the longitudinal drift field. As explained in Section 2.2, the Greinacher voltage multiplier consists of diodes, capacitors and resistors and thus its charging behaviour is that of an RC circuit. Let N be the total number of stages of the multiplier and $U(n,t)$ the output voltage at stage $n \in [0, N]$ after a charging time t, and be $U(n, t=0) = 0 \; \forall n$. Considering a single-stage circuit, i.e. $N = 1$, the output voltage as a function of t is given by the saturation curve

$$U(n=1, t) = U_\infty \left[1 - e^{-t/\tau_1} \right], \qquad (4.1)$$

where U_∞ denotes the output voltage at the last stage of the cascade $n = N$ after an infinite charging time, and is called the setpoint voltage in the following. With R and C being respectively the resistance and the capacitance of the single-stage circuit, $\tau_1 = RC$. Equation 4.1 shows that the setpoint voltage U_∞ is approached asymptotically with t.

Let us now consider a Greinacher multiplier with an arbitrary number N of identical stages and a characteristic time constant τ. The output voltage at stage n of the cascade is expressed by replacing τ_1 in Equation 4.1 by τ_n to take into account all the resistive and capacitive components present from stage 1 to stage n. Since the individual stages are assumed to be designed identically, τ_n can be written as a fraction of τ

$$\tau_n = \tau \frac{n}{N}. \qquad (4.2)$$

As a result,

$$U(n, t) = U_\infty^n \left[1 - e^{-\frac{N}{n\tau} t} \right], \qquad \text{with} \qquad U_\infty^n = U_\infty \frac{n}{N} \qquad (4.3)$$

denoting the setpoint voltage at stage n.

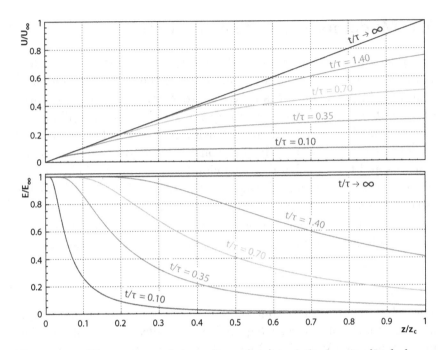

Figure 4.1: Illustrations of the voltage (top) and the longitudinal electric field (bottom, here denoted by E) obtained with the Greinacher multiplier for different charging states t/τ as a function of the drift coordinate z. The graphs result from Equations 4.4 and 4.5. The curves labelled with $t/\tau \to \infty$ correspond to the fully charged Greinacher circuit.

The ARGONTUBE is constructed such that one stage is added to the multiplier cascade for every additional field cage ring electrode (see Section 2.2). Simulations have shown [7] that for the given field cage geometry, the electric field generated in the sensitive volume is uniform. It is thus possible to linearly interpolate the electric potential between two rings along the drift spatial coordinate z and to replace the integer numbers n and N by respectively z and z_c in Equation 4.3, to describe the potential in continuous space. The quantity z_c corresponds to the location of the TPC cathode and $z = 0$ defines the position of the read-out plane which is set to ground potential. The function of the electric potential given in Equation 4.3 is reformulated to

$$U(z,t) = U_\infty \frac{z}{z_c} \left[1 - e^{-\frac{z_c}{z\tau}t} \right], \qquad (4.4)$$

where U_∞ can now be interpreted as the cathode voltage setpoint. Since the read-out plane is connected to ground, $U(z=0,t) = 0 \; \forall t$.

Once the electric potential is known as a function of z, the longitudinal electric field E_L can be calculated by the derivative of $U(z,t)$ with respect to z

$$E_L(z,t) = \frac{\partial U(z,t)}{\partial z} = \frac{U_\infty}{z_c} \left[1 - e^{-\frac{z_c}{z\tau}t} - \frac{z_c t}{z\tau} e^{-\frac{z_c}{z\tau}t} \right]. \qquad (4.5)$$

If $t \to \infty$, the exponential terms in Equations 4.4 and 4.5 vanish and the potential follows a linear course $U(z,t \to \infty) = U_\infty \cdot z/z_c$ with respect to z, corresponding to a constant and uniform drift field $E_L(z,t \to \infty) = U_\infty/z_c$ along z. Obviously, this is the desired situation for operating a TPC. It is clarified by the graphs given in Figure 4.1. The parameter t/τ that appears in the exponential terms of Equations 4.4 and 4.5 is the quantity describing the state to which the circuit is charged. Apart from the situation of a fully charged high voltage multiplier ($t/\tau \to \infty$), four intermediate charging states, reaching from $t/\tau = 0.1$ to $t/\tau = 1.4$, are shown in Figure 4.1. They give an idea of how far the longitudinal electric field is from being uniform when the circuit is not fully charged. In the presence of a leakage current, the fully charged state may only be reached when continuously supplying the circuit. For the ARGONTUBE detector, a small leakage current has been found to be practically unavoidable and, at the same time, charging constantly is not possible either as it induces a strong pickup noise on the sensing wires of the read-out.

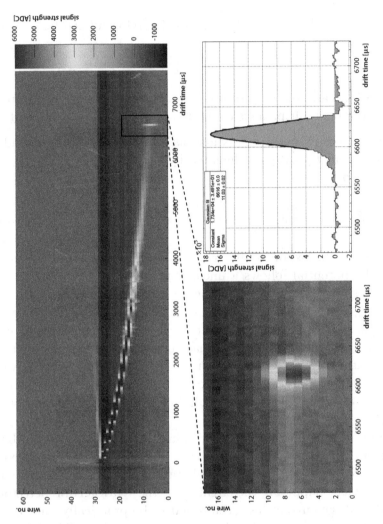

Figure 4.2: *Left:* Drift time measurement with a UV laser beam entering
from the top of the detector, inducing an ionization track
in the sensitive volume and releasing a charge cloud from
the cathode (photoelectric effect, see box). *Bottom right:*
Zoomed view of the charge cloud released from the cathode.
Top right: Gaussian fit of the charge cloud to determine the
drift time that corresponds to the position of the cathode
and to the full length of the TPC.

4.2 The electric field in ARGONTUBE

One example of a laser track recorded during the ARGONTUBE run in October 2013 is shown in Figure 4.2 for a cathode voltage setpoint of $U_\infty = 130.0 \pm 2.0\,\mathrm{kV}$. The laser beam enters from the top of the detector with a certain angle, passes through the argon volume, where it leaves a straight ionization track, and finally hits the cathode where a larger amount of charge is released from the metal due to the photo-electric effect (see Fig. 4.2, box). A zoomed view of the charge cloud originating from the cathode is shown together with the histogram of the projection of wires 4 to 11. The endpoint is fitted with a Gaussian. The total drift time is obtained from the fit and is of the order of several ms (here $t_{\mathrm{tot}} = 6.616\,\mathrm{ms}$, with a negligible measurement uncertainty) at typical field strengths achieved in ARGONTUBE. The recorded ionization track clearly shows deviations from a straight line and thus indicates that the electric field is non-uniform. A priori, one must assume that the distortions are due to both, longitudinal disuniformities and a parasitic transverse component of the electric field.

The model in Equation 4.5 describes E_L as a function of (z, t) and contains the parameters U_∞, z_c and t/τ. U_∞ and z_c are known for ARGONTUBE for a given setpoint voltage. The charging state of the Greinacher circuit, i.e. t/τ, however, is unknown and must be treated as a free parameter. The drift time $t_d(z)$ needed for a test charge released at the drift coordinate z to reach the read-out plane ($z = 0$), is obtained by an integration along the drift direction

$$t_d(z) = \int_0^z \frac{1}{v_L(E_L(z'))}\, dz', \qquad (4.6)$$

where $v_L\left(E_L\left(z'\right)\right)$ denotes the longitudinal drift velocity in the interval $[z', z' + dz']$. The mobility and thus the drift velocity of the electrons depends in fact on the total electric field strength rather than just the longitudinal or the transverse component. However, as will be obtained below, E_T is small compared to E_L (order of less than

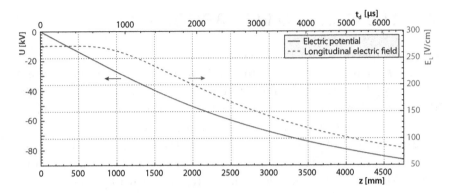

Figure 4.3: Using the model described by Equation 4.4 and the measured full-length drift time, the electric potential (solid line,
left y-axis) and the longitudinal electric field E_L (dashed
line, right y-axis) along the z coordinate can be derived by
means of Equation 4.6.

$2\,\%$) and one can set $E_{\text{tot}} \approx E_L$. Hence, $v_L = v_L(E_L)$ is a good
approximation. The relationship between v_L and E_L is well-known
from several experiments [38–41].

The formula for the drift time given in Equation 4.6 depends through
E_L on the yet unknown parameter t/τ. By using a full-length laser
track as shown in Figure 4.2, the total drift time t_{tot} across the drift
gap can be measured and one can set the constraint $t_d(z_c) = t_{\text{tot}}$,
allowing the calculation of t/τ. With a cathode voltage setpoint value
of $U_\infty = 130.0 \pm 2.0\,\text{kV}$, a total drift length of $z_c = 4.76\,\text{m}$, and
the total drift time $t_{\text{tot}} = 6.616\,\text{ms}$, obtained from the laser track
in Figure 4.2, one determines $t/\tau = 1.09424 \pm 0.00002$ (compare
Fig. 4.1). Hence, the Greinacher circuit was far from fully charged
when these tracks were recorded, even though it had been charged
until a stable minimum charging current was reached. The situation
that corresponds to the actual quoted value of t/τ is presented in
Figure 4.3 where $U(z)$ and $E_L(z)$ are given by respectively the solid
and the dashed curves. The plot features two different scales along the
abscissa, drift time t_d (top axis) and drift coordinate z (bottom axis),

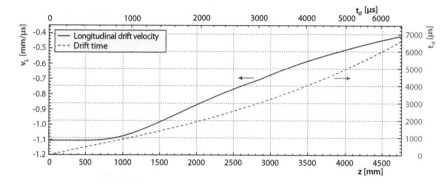

Figure 4.4: Once E_L is known, the longitudinal drift velocity v_L (solid line, left y-axis) can be obtained using the electron mobility [39]. The drift time (dashed line, right y-axis) is calculated by means of Equation 4.6.

which are related by the function $t_d(z)$. The longitudinal electric field strength is a monotonically decreasing function and reaches a value of about $1/3\,E_L(z\!=\!0)$ at the cathode. For illustrative purposes, $v_L(z)$ (solid line) and $t_d(z)$ (dashed line) are shown in Figure 4.4. Note that $t_d(z = z_c) = 6.616\,\text{ms}$ which marks the constraint that was used to determine the parameter t/τ.

Using the analytic function in Equation 4.5 describing $E_L(z)$ and the measured parameter t/τ, one can correct the rawdata UV laser induced track for the longitudinal disuniformities. When applying this correction to the data shown in Figure 4.2, a track of much reduced curvature than the original one results. By slicing the corrected ionization track along the drift coordinate and determining the center of gravity in the x coordinate for each slice $[z, z + \Delta z]$, the graph given in Figure 4.5 (top) is obtained. It corresponds to the reconstructed collection view of the laser track corrected for longitudinal electric field disuniformities. The wire number is translated into the x coordinate by use of the ARGONTUBE wire spacing of 3 mm. Even though correcting for E_L reduces the curvature of the track, there is still a knee clearly visible at about $z = 700\,\text{mm}$. The residual deviations of the track

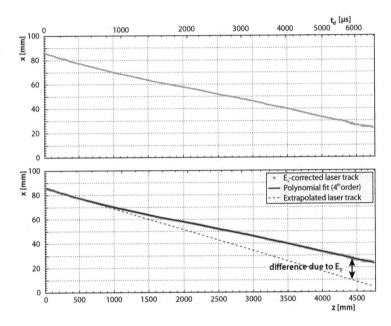

Figure 4.5: *Top:* Measured laser ionization track corrected for the longitudinal disuniformities. Error bars are not shown for reasons
of visibility. *Bottom:* Laser track fitted with a 4th-order
polynomial (solid line) to smoothen the measurement for
reasons discussed in the text. The dashed line represents
the deduced actual laser track. If follows from a fit to the
first segment of the track corrected for $E_L(z)$ (measurement
points) assuming that $E_T = 0$ at the wireplane $(z = 0)$.

corrected for E_L from a straight line are fully attributed to the
parasitic transverse electric field present in the detector along the
laser beam path. To find an estimate for the strength of E_T, the
assumption is made that there is no transverse component at the
location of the wireplane, i.e. $E_T(z = 0) = 0$. With that in mind,
the effective path of the laser beam can be reconstructed by fitting a
straight line (dashed) to the first straight segment of the measurement
points in Figure 4.5 (top). This line is referred to as the deduced
actual laser track and is expected to correspond to the actual laser

beam path. Furthermore, the measurement points are fitted with a polynomial of degree four (solid line) in order to limit the degrees of freedom of the data points and to obtain a smooth curve. This is of importance in the following to avoid numerical issues as the variations from point to point, caused by the detector resolution and noise, would lead to meaningless results for E_T. The difference in x between the straight line fit and the fourth order polynomial at each z is taken. This yields the displacement $\Delta x(z)$ of the deduced actual laser track, caused by the parasitic transverse field component $E_T(z)$. $\Delta x(z)$ is represented by the solid line in Figure 4.6. The transverse drift velocity v_T (dashed line) is deduced from the derivative of Δx with respect to t_d. To obtain meaningful results, it must be ensured that there is a smooth transition from the boundary condition $\Delta x(z\!=\!0) = 0$ to the values at $z > 0$. Once $v_T(z)$ is known, it is possible to determine $E_T(z)$. To perform this step, the electron mobility $\mu(z)$ must be calculated first using $E_{\text{tot}}(z) \approx E_L(z)$. Subsequently, $E_T(z)$ follows from $E_T(z) = v_T(z)/\mu(z)$.

Both components of the electric field are shown together in Figure 4.7 on top. E_T is in general by about two orders of magnitude smaller than E_L and the ratio of E_T/E_L (bottom) is at maximum at z_c where it amounts to about 1.6%. While the longitudinal field component is valid for the whole sensitive volume of the detector, the transverse component is only valid along the laser path of the specific track and was found to vary within the detector volume. The example track shown here was found to be the one leading to the largest E_T, given the limitations in angle and position to steer the laser beam into the ARGONTUBE (compare Sec. 2.4). From studying cosmic muon tracks, one expects the transverse component to be even larger in certain regions of the sensitive volume, for instance in proximity of the Greinacher circuit. Moreover, only one projection (the collection view) was considered here and more information about the electric field could be obtained by also taking into account the induction view.

To check the consistency of the electric field model, the measurements in Figure 4.5 are again considered. Since the charge cloud released

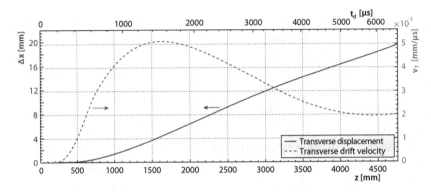

Figure 4.6: The solid line shows the transverse displacement Δx caused by the parasitic electric field E_T. It results from the difference between the straight line fit and the 4th-order polynomial in Figure 4.5. Considering Δx and the drift time for each z, one can determine the transverse drift velocity v_T at each point along the drift coordinate (dashed line).

from the cathode is fully contained for the given example track (see Fig. 4.2, bottom left), the straight line fit in Figure 4.5 (dashed line), which represents the deduced actual laser path, must stay and hit the cathode within the sensitive volume. The beam center x_{BC} must hence be at an $x(z_c)$ coordinate larger than $w/2$, where w denotes the width in x within which the beam spot releases photoelectrons. A first order estimate of the beam spot size at the cathode, located at roughly 10 m from the laser source, taking into account only the specified beam divergence of 0.6 mrad, gives a value of about 6 mm. By using the width of the spot in Figure 4.2 on the bottom left, one can find a second, independent estimate. A Gaussian fit to the measured endpoint yields a transverse width of about $\sigma \approx 5$ mm. This value can be corrected for transverse diffusion for the given drift time. When using a transverse diffusion coefficient of $D_T = 4.2 \pm 0.4 \, \text{cm}^2/\text{s}$ [8], the corrected full width at half maximum of the beam at the cathode amounts to about 7 mm, consistent with the first estimate. For the track corrected for E_L presented in Figure 4.5, the linear fit on the first straight segment yields

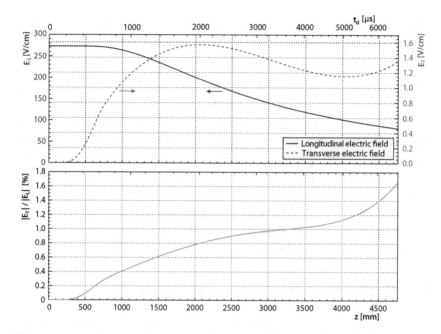

Figure 4.7: *Top:* Summary of the two electric field components as obtained by the procedure explained in the text. *Bottom:* Ratio of the absolute transverse to the longitudinal field strengths along the drift direction (in percent).

$x_{BC}(z) = (86.12 \pm 0.20\,\mathrm{mm}) - (0.0177 \pm 0.0011) \cdot z$ and consequently $x_{BC}(z_c) = 2.0 \pm 5.1\,\mathrm{mm}$, compatible with a fully contained endpoint. By looking at the position of the charge cloud in Figure 4.2 on the bottom left (wire no. 7 or $x \approx 24\,\mathrm{mm}$) and considering that the parasitic transverse field has moved the charge cloud towards larger x during the drifting process, one finds that the actual beam spot must have indeed been close to the border of the sensitive volume. For a second measurement, the laser beam was slightly retuned such that the endpoint moved just out of the sensitive volume and was not visible anymore in the acquired data. The same correction for E_L was applied as in the previous case and the first segment of the laser track corrected for E_L was again fitted with a straight line. This time, one would expect x_{BC} to be outside of the sensitive volume, i.e. $x_{BC} < -w/2$. Indeed, $x_{BC}(z) = (84.98 \pm 0.21)$ mm $- (0.0193 \pm 0.0011) \cdot z$ and hence $x_{BC}(z_c) = -6.8 \pm 5.0$ mm, consistent with the expectations.

Although the model for the longitudinal electric field introduced here needs further validation, these measurements already prove a certain degree of reliability. To improve the measurements and to fully verify the model, the direction and position of the UV laser beam entering the detector should be measured and compared to the results obtained from the deduced actual laser track and to the position x_{BC} where the beam hits the TPC cathode. From cosmic muon studies, presented in Chapter 6, the here described model was used and results were obtained that are consistent with values reported from other, independent experiments.

4.3 Cathode voltage measurements

The absolute value of the electric potential should increase linearly towards the TPC cathode to guarantee a uniform electric field in the detector. As discussed in the previous sections, UV laser induced tracks combined with the model for the Greinacher circuit indicate that there is a significant decrease in the longitudinal electric field strength with increasing drift coordinate z in ARGONTUBE. Hence, towards the cathode, the absolute voltages that are reached deviate more and more from the setpoints. To measure the cathode voltage, an electric field mill was installed on the bottom flange of the inner cryostat vessel (see Fig. 4.8, left), about 30 cm below the cathode plate. This device was chosen since the cathode potential in ARGONTUBE cannot be measured by standard voltmeters. They would obtain the voltage from a current flowing through a resistor from cathode to ground and hence quickly discharge the Greinacher circuit. A field mill, on the other hand, is able to measure the strength of an electric field without actually affecting it. A rotor with cut out windows rotates above sensor electrodes and periodically shields and unshields them from the electric field. While uncovered, the presence of the electric field causes induction of charge on the sensor electrodes. When covered and shielded, the sensor electrodes lose their charge by a current flow through a resistor to ground. The strength of the current can be determined by measuring the voltage drop across the resistor. The periodic charging and discharging of the electrodes is registered as a sinusoidal variation of this voltage. Its frequency is proportional to the rotation frequency of the rotor and the amplitude is proportional to the induced charge which is related to the electric field strength. When measuring the signal generated by the field mill for a number of different and known cathode potentials, this instrument can be calibrated which makes the measurement of the cathode voltage possible. Such a calibration has been done in air, inside the ARGONTUBE cryostat with the fully assembled TPC in place and the field mill installed on the bottom flange below the cathode. The voltage was generated by a high voltage power supply (Spellman SL130-150) and brought to the cathode by means of a high voltage coaxial cable lowered into the detector. For a number of voltages ranging from zero to 48 kV, the signal amplitude of the field mill was measured. The upper limit of 48 kV was given by the dielectric strength of air and the distance from the cathode to the cryostat wall.

The solid line in Figure 4.9 represents the absolute cathode voltage measured with the field mill during detector operation versus the value expected for a fully charged Greinacher cascade at the given peak-to-peak input voltage. To obtain the measurement values, the field mill calibration curve in air was translated to the situation of the cryostat filled with liquid argon simply by using the ratio of the dielectric constants of air and of liquid argon (see Tab. 1.1). The dependence of the measured voltage on the Greinacher setpoint voltage is linear and the measured value at zero input voltage is compatible with zero. The error bars are propagated from the calibration curve, measurement errors of the field mill signal and the uncertainty in the dielectric constant of liquid argon.

The validity of the field mill measurements is highly questionable, mainly for the reason that the calibration in air was translated to

Figure 4.8: *Left:* The field mill is installed on the bottom flange of the inner vessel. The sensor electrodes and the rotor are both visible. The rotor position corresponds to fully unshielded sensors (compare text). *Right:* CAD drawing of the lowermost part of the TPC.

the real situation with liquid argon simply by means of its dielectric constant. This procedure would be correct only if the volume between the cathode and the field mill was filled entirely with liquid argon. In any other case, the change of the electric field strength caused by the change of the dielectric properties of the materials in the space between the field mill and the cathode (e.g. as a result of the much lower temperature) as well as their geometries have to be taken into account. In fact, in ARGONTUBE there is a structure made of PAI installed between the field mill and the cathode which serves as a spacer for the TPC to keep it well centered inside the inner vessel (see Fig. 4.8, right). To the author's knowledge, the dielectric constant of PAI has not yet been measured at liquid argon temperature. However, when lowering the temperature from $25\,^{\circ}\mathrm{C}$ to $-186\,^{\circ}\mathrm{C}$, a decrease in the dielectric constant is expected for polymeric dielectric materials [83]. Indeed, measurements made for PAI in a range of temperatures down to $-100\,^{\circ}\mathrm{C}$ indicate a significant decrease in its dielectric constant [84]. Consequently, concerning the ARGONTUBE field mill measurements, the way how the calibration in air was adapted to the situation with liquid argon would be incorrect and taking into account the decrease

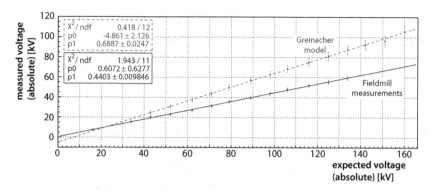

Figure 4.9: Field mill measurements (solid line) and Greinacher model calculations (dashed line, see Eq. 4.4) of the absolute cathode voltage versus the expected one, derived from the Greinacher peak-to-peak input voltage multiplied by the number of cascade stages. A large discrepancy is observed between the two methods. The Greinacher model calculations are considered more reliable for reasons discussed in the text.

in the dielectric constant of the PAI structure, the absolute cathode voltages would be shifted to larger values than those reported in Figure 4.9. However, knowledge of the dielectric constant of PAI alone is not enough to correctly adapt the calibration done in air to the real situation with liquid argon. The change in the electric field strength would have to be simulated for the geometry of the PAI structure and the liquid argon volume as it is in ARGONTUBE. A more straightforward approach, although not free from technical difficulties, would be to perform the calibration directly in liquid argon instead of air.

The Greinacher model, introduced earlier in this chapter, combined with UV laser induced events was used as an alternative method to measure the cathode voltage in ARGONTUBE. The results are given by the dashed line in Figure 4.9. The error bars originate from an uncertainty in the fit parameter t/τ, the full drift length z_c and from an uncertainty in the peak-to-peak input voltage of the Greinacher circuit. Again, a linear dependence is clearly visible, however, there is a slight incompatibility with the voltage at zero input. The values obtained with the Greinacher model strongly differ from the field mill measurements. The Greinacher model is well understood and has been proven to be consistent by means of UV laser induced events as well as by the results obtained in Chapter 6. The method employing the field mill on the other hand, includes an unknown effect originating from the dielectric properties of PAI at liquid argon temperature. The strength of the decrease in the dielectric constant of PAI is of an order that could explain the discrepancy between the field mill measurements and the values obtained from the Greinacher model. This is, however, yet to be studied in more detail. At the present moment, the cathode voltage measurements obtained by means of the Greinacher model are considered more reliable.

5 Realization of a GPU-based track finder

To process efficiently the ARGONTUBE cosmic muon events collected during the runs in June and October 2013, a track finder based on the Hough transform was implemented in CERN ROOT and compute-intensive parts of the code were realized on a NVIDIA graphics processing unit (GPU) using the CUDA (Compute Unified Device Architecture)[1] C language. The Hough transform is a pattern recognition algorithm and can be used to detect any arbitrary shape in an image. In this thesis, the PCLines algorithm [85–87] has been used to detect segments of straight lines in ARGONTUBE data for the identification of particle ionization tracks. It is a computationally fast, modified Hough transform and especially well suited to be run on GPUs. The here described track finder may be used as a first stage to generate a seed (initial state) for more sophisticated tracking algorithms based on Kalman filtering methods [9, 88] which are implemented for example in the LArSoft[2] framework [88]. LArSoft (Liquid Argon Software) is a project devoted to the reconstruction of LArTPC data and reliable detector simulation. It is a flexible toolkit which is adaptable to TPCs with different geometries and readouts.

5.1 The Hough transform

To find specific objects or shapes (e.g. lines, circles, rectangles, ellipses) in raster images, so called pattern recognition algorithms are used.

---

[1] developer.nvidia.com/cuda
[2] https://cdcvs.fnal.gov/redmine/projects/larsoftsvn

A prominent and powerful one is the Hough transform which was originally developed to automatize the analysis of bubble chamber pictures [89]. At that time it was used to search for straight lines. Later, it was extended to any parameterizable objects [90] and finally generalized to arbitrary shapes [91]. In the following, the general idea of the Hough algorithm for the detection of parameterizable objects is illustrated on the basis of straight lines.

A straight line in a 2D coordinate space x-y is unambiguously described by a set of two parameters (a, b). There are different ways to choose a and b, however, in the most common parameterization they describe the slope of the line and its intercept with the y axis respectively. A line in x-y space would thus be represented by $y(x) = ax + b$. The idea of the Hough transform is to generate a new parameter space, named the Hough space, which is spanned by the parameters describing the object of interest. For the straight line, they are a and b (see Fig. 5.1). Each pixel of the image has a certain intensity (weight). In Figure 5.1 on the left it is either 1 (white) or 0 (black). By means of the coordinate transform

$$y(x) = \mathbf{a}x + \mathbf{b} \quad \Leftrightarrow \quad \mathbf{b}(a) = y - \mathbf{a}x, \tag{5.1}$$

the pixel (x_i, y_i) of the image space is transformed to the Hough space, depicted in Figure 5.1 on the right, where it 'votes' according to its intensity. In the Hough space, $b(a) = y_i - ax_i$ again describes a line which represents all the possible lines in the image space passing through the point (x_i, y_i). As a result, the transform in Equation 5.1 describes a point-to-line mapping (PTLM), meaning that a point in image space is mapped to a line in Hough space and vice versa.

When applying the transform given above to every pixel of an image containing pixels of different intensities, and weighting for each pixel with its intensity (voting process), certain locations in the Hough space will get more votes than others as all the points lying on one line in the image space will agree on the same parameter values. The local maxima in the Hough space thus represent the object of interest

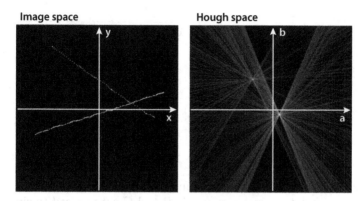

Figure 5.1: Example of the Hough transform for straight lines with the slope-intercept parameterization. The images were generated with a Hough transform java applet [92].

candidates as illustrated in Figure 5.1 for two different lines. This concept applies identically to any other parameterizable object. To summarize, the Hough transform reduces the pattern recognition problem in the image space to a maximum finding problem in the Hough space.

The method offers a number of advantages. It is relatively stable against noise which is a very important aspect when analyzing real detector data. This originates from the fact that a noisy pixel in the image will generally contribute only with a low weight in the Hough space. Moreover, random noise in the image will lead to an overall background in the Hough space rather than to specific maxima. Often, most of the noise is cut away already in the image before by means of thresholding or edge-detection algorithms [93] that isolate the pixels corresponding to structure in the image. A second advantage is robustness against gaps in the pattern. The line with the negative slope in Figure 5.1 is composed of points that are only sparsely distributed along it. Nevertheless, a local maximum is clearly established in the second quadrant of the Hough space. Last but not least, the parallelization of the transform is straightforward as every pixel of

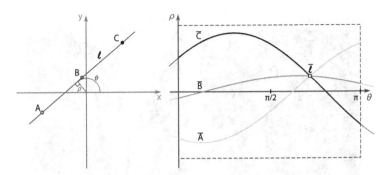

Figure 5.2: The Hough transform in polar coordinates (ρ, θ) [85]. The points in image space A, B and C are mapped to sinusoidal lines $\overline{A}$, $\overline{B}$ and $\overline{C}$ in Hough space, whereas the line l is represented by a point $\overline{l}$.

the image can be transformed independently of the others. These are ideal prerequisites for the use of GPUs.

The Hough method also has a number of drawbacks, though. First of all, an actual implementation must use a discretized Hough space (the so called accumulator array) rather than a continuous one. This means that similar to the input image, the Hough space is rasterized into pixels. Additionally, the accumulator array cannot have an infinite number of entries, which is why the slope-intercept parameterization introduced above (see Eq. 5.1) is not usually used. After all, in this form vertical lines have an infinite slope and thus the parameter space is unbounded, even for images of a finite size. However, it has been proven [94] that for any PTLM a complementary PTLM can be found such that these two mappings define two finite Hough spaces that include all the possible lines of the image space. Another possibility to tackle this problem is by introducing polar coordinates (ρ, θ) [90], which is a commonly used parameterization nowadays (see Fig. 5.2). In this form, a point in image space is mapped to a sinusoidal line in the Hough space which is thus bounded for bounded images. On the other hand, in terms of computing speed they are disfavoured for their use of trigonometric functions. Many other parameterizations

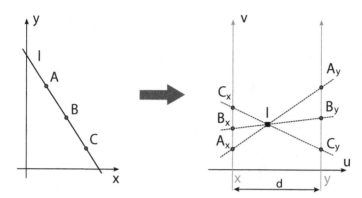

Figure 5.3: The use of parallel coordinates for the description of lines in 2D [86].

are available for lines such as the Fan-beam (bounding circle), Muff (bounding rectangle) or the circle transform. They are discussed and compared to each other along with their advantages and disadvantages in [85]. For the ARGONTUBE track finder, several of these parameterizations were considered and implemented for comparison.

The discretization of the Hough space introduces two issues – resolution limitation and peak splitting. In theory, the resolution can be increased by finer segmentation of the parameter space. It would only be a question of available amount of memory to store such a large array of data. However, apart from the memory issue, a finer segmentation also leads to a distribution of the votes attributed to one track in the image into several bins in the parameter space and thus reduces the intensity of local maxima, especially in the presence of noise and for tracks that do not follow a perfectly straight line (compare Sec. 5.3). The peak splitting or vote spreading problem is discussed extensively in [95] and results from a slight mismatch between the parameter values of a line in image space with the binning in the Hough space. Reference [96] suggests a technique on how to successfully solve it. Another drawback is encountered when one uses the Hough method for the detection of straight segments in an image. The positions

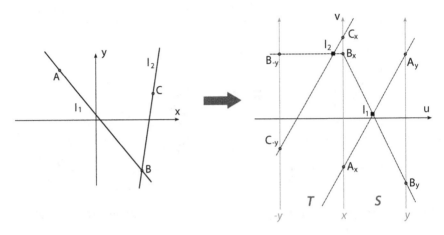

Figure 5.4: Example of two lines I_1 and I_2 situated in the straight S
and in the twisted T spaces respectively when represented
in parallel coordinates [86]. T and S are complementary and
contain all the possible lines of the image space.

of the local maxima in the parameter space only give information
about the line containing the segment but not about the position and
length of the segment itself. Studies exist which suggest to exploit
better the information contained in the Hough space, such as the
butterfly-shaped distribution of votes (see Fig. 5.1, right) around local
maxima [97].

A parameterization of particular interest, which was eventually used
to realize the ARGONTUBE track finder, are parallel coordinates (PC),
or, the so called PCLines algorithm described in much detail in [85–
87]. It has been shown that this approach is more accurate than
existing ones, computationally extremely efficient, as it does not use
any trigonometric functions, and that it can be directly implemented
on the GPU. Peak splitting and resolution limitation are, however,
existing issues also for the PCLines algorithm. Usually, PC are used
to visualize high-dimensional data. The principle is illustrated in
Figure 5.3 for a 2D coordinate system x-y. The PC space is spanned
by the x and y coordinate axes both set vertically (in green) and

separated by a distance d from each other along the abscissa labelled u. In this frame, the point A is represented by its two coordinates A_x and A_y, visualized by the dots printed on the x and y axis respectively and by the dashed line that connects them. B and C can be visualized in the same manner. The three points A, B and C are located on one straight line in the x-y space and, as a result, the dashed lines in the PC space meet in one point labelled I. In fact, PC are another form of PTLM, i.e. a point in PC is mapped to a line in the x-y space and vice versa. As mentioned above, an issue of PTLM is that their parameter space is unbounded meaning that the PC system given in Figure 5.3 on the right does not include all the lines one can think of in the bounded x-y reference frame. By introducing a complementary space called the twisted space T, which is given by the x and the reverted y axes in addition to the already present straight space S, spanned by the parallel x and y axes, all the possible lines of the image space are included in the PC reference frame. Figure 5.4 gives an example with two lines I_1 and I_2 which are situated in the straight and in the twisted space respectively.

5.2 Brief introduction to CUDA GPGPU programming

The parallel computing platform CUDA was initiated by NVIDIA [98] and offers a convenient and accessible way to general-purpose GPU programming (GPGPU). It has gathered a lot of attention in many fields of science and in the industry[3] over the past few years. The idea of GPGPU programming is to use the abilities of GPUs not for computer graphics, which is what they were originally made for, but for any computing tasks that can benefit from their massively parallel computing architecture. Other than the CPU (central processing unit), a GPU includes up to several thousands of individual computing cores at the expenses of less sophisticated flow control and data caching. It is optimized for an extremely fast execution of

---

[3]developer.nvidia.com/cuda-action-research-apps

massively parallel code where a mathematical operation is applied to a large set of data simultaneously. The idea is to use calculations rather than big data caches to reduce the time penalties originating from memory access latencies. GPUs are thus especially fast for computational problems with a high arithmetic intensity, that is the ratio of arithmetic operations to memory operations is very large. The speed-up compared to CPUs for such specific computations can be up to several orders of magnitude. A computational problem which is ideally suited for GPUs is the multiplication of two large matrices. Since all the entries of the output matrix are calculated independently of one another, their computation can be distributed to different and independent computing cores running the calculations simultaneously.

To process data on a NVIDIA GPU (device), a so called *kernel* function needs to be implemented. It is written in the CUDA C language and contains the algorithm to calculate the output data on the GPU. An important limitation for GPGPU programming is that there is no possibility for the GPU to access the host memory (main RAM) directly. Similarly, the CPU is not able to dereference a memory location on the graphics card. As a result, before calling the GPU kernel, all the input data that are to be processed by the GPU must be copied to device memory. The memory allocation on the device for input and output data must be taken care of by the programmer. Once the input data are copied to device memory, the kernel is called by the CPU (host). With the kernel call, the configuration of how to distribute the computational tasks among the individual cores of the GPU, i.e. the thread configuration, is passed to the device. The kernel call is done asynchronously meaning that the host can execute further operations while the kernel is still running on the device. In practice, however, the host must usually wait for the output data from the GPU computation and is thus idle in the meantime. After completion of the kernel, the output data are copied back to the host memory and may be processed further. It is worth noting that for certain computational tasks, the overhead of copying the data from host to device and back can represent a serious bottleneck.

Programming a kernel function is a rather delicate task, especially when it comes to optimizations that need a full understanding of the device architecture and its memory management. A detailed discussion of these topics is out of the scope of this thesis and the reader is referred to specific literature [98, 99]. An important aspect of parallelized code is thread-safety, meaning that race-conditions must be avoided and thread synchronization has to be taken care of. A race-condition occurs when the output of a function executed in parallel threads is not determined, but depends on the sequence of execution of threads and memory accesses. This may happen when the synchronization of threads is not done properly. In CUDA, the latter issue is resolved by using the synchronization function, called at the right places in the code. Furthermore, CUDA allows the use of so called atomic operations which means that the first thread accessing a certain location in memory locks it, performs its operations and unlocks it afterwards. During the operations no other thread is permitted access. This kind of operation must be used with care. If the kernel consists of operations where many different threads are about to access the same location in memory via atomic operations, its execution will be extremely slow. As an example, the voting process, which is a core part of the Hough transform, is a task where different threads may want to access the same location in memory. For the ARGONTUBE track finder, the atomic operations provided by CUDA were used to guarantee thread-safety.

5.3 The GPU-based track finder for ARGONTUBE

Compared to other detector technologies where an ionization track is composed of individual and well separated hits (e.g. nuclear emulsions), LArTPCs offer the advantage of producing data with well connected tracks, meaning that the individual hits overlap with their neighbours. By use of this criterion a simple and fast track finder could be realized. However, in the case where ionization tracks include

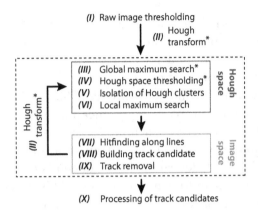

Figure 5.5: Schematic of the track finding procedure. Tasks denoted with the superscript (*) are those running on the GPU.

gaps, e.g. due to dead read-out channels, or for events where two or more tracks cross, this simple tracking algorithm has to be extended to be able to successfully find the corresponding hits of a specific track or to separate individual tracks respectively. One possible and well suited approach is to use a pattern recognition algorithm such as the Hough transform. The obvious choice for the kind of shape to be recognized in ARGONTUBE events are straight lines, since the ionization tracks are composed of piecewise straight segments. The PCLines algorithm offers advantageous properties in terms of computing speed and accuracy for the detection of this kind of shape [86, 87] and was thus used for the implementation of the ARGONTUBE track finder.

The schematic in Figure 5.5 gives an overview on the main parts of the reconstruction tool. It is based on an iterative approach, indicated by the tasks contained in the dashed rectangle, meaning that individual track candidates in an event are identified one after the other. The working principle is illustrated by means of an example event shown in Figure 5.6 (a). The event contains two ionization tracks induced by two cosmic muons and was recorded during the ARGONTUBE run in June 2013. The image shows rawdata and the abscissa is given in ADC time samples in units of bins, each having a width of 1.01 μs.

In this scale, the spatial location of the wireplane corresponds to the time at bin 590. Hence, the track entering from the left passed through the wireplane and crossed a large part of the sensitive volume. The position of the TPC cathode corresponds to the time at bin 7720 which is not within the range of the figure.

To reconstruct the two tracks, the first step is to do a thresholding (I) of the raw data event (see Fig. 5.6, a) which separates the pixels containing signal from those containing noise. Pixels exceeding the threshold are Hough transformed by the GPU (II). The threshold above which a pixel is no more considered to be noise is chosen such as to include the large majority of true hits. As a result, many noisy pixels are transformed as well. Although the Hough transform is relatively robust against noise, a larger number of pixels to be transformed results in a performance loss of the Hough algorithm. A more sophisticated approach to identify pixels containing signal would be an edge-detection algorithm [93]. An extract of the Hough space resulting from the event in Figure 5.6 (a) is shown in Figure 5.6 (b). As expected from the two tracks, two local maxima appear in the Hough space one of which is marked by the rectangle in the figure. The second maximum is located at the Hough coordinates ($u = 500, v = 260$) and corresponds to the second track. The clusters are very well separated due to the different inclination and position of the tracks in the image space. By identifying and isolating the two maxima in the Hough space, the two crossing tracks can be separated. However, the Hough maxima are not pointlike but largely extended, which is due to the fact that the recorded ionization tracks have a curvature. It results from particle scattering with the argon atoms and from electric field disuniformities inside the drift volume. The latter is considered the major contribution in the case of ARGONTUBE (see Chapter 4).

To isolate the individual maxima (called 'clusters' in the following) a second thresholding process is performed, this time in the Hough space (IV). For this task, each v coordinate is independently searched for connected, i.e. neighbouring, bins above a certain threshold, determined from the global maximum of the Hough space (III). Finding

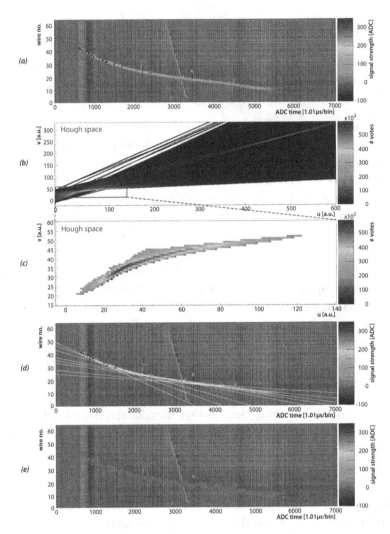

Figure 5.6: (a) Example of an event recorded with ARGONTUBE containing two crossing muon tracks. (b) The Hough space obtained from tasks (I) and (II) contains the two distinct maxima at $(u_1 = 40, v_1 = 35)$ and $(u_2 = 500, v_2 = 260)$ originating from the two tracks. (c) Zoomed view of one of the clusters, isolated after performing (III) to (V). (d) Lines that were found in the local maximum search (VI). (e) Event after removal of the identified track candidate by carrying out (VII) to (IX).

the global maximum, and thresholding in the Hough space are well suited tasks to be executed on the GPU. Finding a maximum in a large array can be done efficiently by means of a standard reduction algorithm [100]. Since the thresholding operation is done independently for each value of v, it can be distributed among the GPU cores. In the next step, the individual series of connected bins at every v coordinate are compared for overlaps with those at the neighbouring coordinates $v + 1$ and $v - 1$. By performing this operation for any of the connected bin series found during the thresholding process (IV), the Hough clusters are built (V). One example of an isolated Hough cluster is shown in Figure 5.6 (c). Once the clusters are known, the one having the most votes is selected and a local maximum search is done within its area (VI). This search is based on a standard approach, where the intensity of each pixel is compared to those within a certain neighbourhood. The locations of these maxima represent different straight lines in the image space. For each maximum found, one line is drawn in the image space (see Fig. 5.6, d), and along each of the lines, hits are searched for (VII) and finally structured together to form a track candidate (VIII). The hits have to fulfill certain criteria such as not to be isolated and on their own, but to have at least one other hit connected to them. At the end of an iteration, the track candidate is removed from the image space (IX) as illustrated in Figure 5.6 (e). Tasks (II) to (IX) are repeated until the Hough space global maximum falls below a certain threshold derived from 'empty' events, i.e. events that contain only noise.

The idea of the iterative approach is illustrated in Figure 5.7 which shows the continuation of the reconstruction of the event given in Figure 5.6. The first image (Fig. 5.7, a) depicts the Hough space for the case where the first track candidate was removed from the image space. The cluster that was located at ($u = 40, v = 35$) is no more present. But, the cluster belonging to the second track is now clearly identified. A zoomed view, after performing steps (III) to (V), is shown in Figure 5.7 (b). The lines corresponding to the local maxima are given in subfigure (c). The final picture (d) shows a

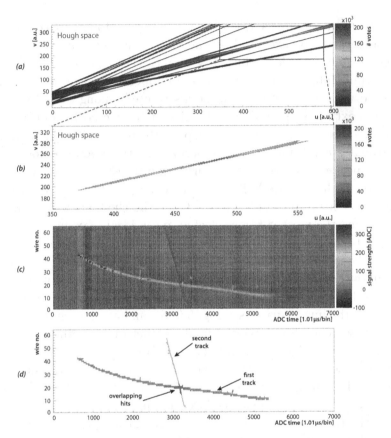

Figure 5.7: Second and last iteration of the algorithm for the given
event. (a) Hough space of the image shown in Figure 5.6 (e).
(b) Zoomed view of the second isolated cluster. (c) Lines
that were found in the local maximum search. (d) Masks
of the two reconstructed tracks. Several bins at the crossing point are attributed to both tracks and need further
processing.

mask of the two tracks that were identified. At the crossing point of
the two reconstructed track candidates, several bins are allocated to
both tracks and must be processed further (X). In a first approach,

this could be done for example by interpolating each of the crossing tracks using their signal width and strength to determine the fraction of charge that must be attributed to each of them. For events with clearly separated Hough clusters, such as for the example shown in Figures 5.6 and 5.7, the iterative approach may seem unnecessarily complicated and slows down the reconstruction. In a first version of the track finder, the Hough transform was carried out only once and all the clusters were isolated simultaneously. This method had a number of drawbacks, though. One of them is that the Hough space threshold must be chosen very low to also successfully find the 'weak' clusters, corresponding to short, weak or strongly curved tracks. A low threshold, however, automatically leads to largely extended Hough clusters during steps (IV) and (V) (see Fig. 5.5), which are no more clearly separated for certain events. Moreover, it is not a priori clear how low the threshold must be set. In case it is not chosen low enough, short and weak tracks may not be reconstructed. The iterative method, on the other hand, is beneficial as soon as there are several clusters lying close to each other in Hough space. In every iteration, the global maximum in the Hough space is determined first (III) and the threshold which is needed to perform task (IV) is adapted accordingly. By using a strict limit on the threshold, only the central part of a Hough cluster is isolated and it is thus clearly separated from the neighbouring clusters. Also, it turned out that using only the local maxima within the central area of the Hough cluster is enough to search for hits along the lines in the image space (VII) and to find the entire track candidate. Finally, since the Hough transform can be sped up by means of the GPU, as will be discussed in the following section, the iterative method still represents a time efficient solution.

5.4 Computational speed-up of the Hough transform

To quantify the computational speed-up that is achieved when performing the Hough transform on the GPU instead of the CPU (serial code), the time needed to calculate the Hough accumulator array was measured as a function of the number of pixels that are transformed. Note that for this measurement, only the Hough transform itself was considered rather than the total time per event needed for the entire track finding procedure. Real ARGONTUBE events were processed, using the same thresholds as for reconstruction, to get a descriptive value of the Hough transform speed-up. For the GPU time measurement, using real data is essential to obtain meaningful values. With input data that were simulated by perfectly straight lines, or by only one pixel transformed many times, the GPU code would not execute as fast since different threads would more often want to access simultaneously the same indices in the Hough accumulator array, i.e. the same address in memory. On the other hand, a situation with pixels above threshold randomly distributed in the image, would most likely end up in the fastest execution of the GPU code. Concerning the serial code on the CPU, there are no such memory access issues as there is only one thread running at a time and the pixels are transformed and the Hough accumulator array filled one after the other.

The measurements were performed on a system with an Intel Core i7-860 CPU with 4 GB of RAM (1333 MHz) and an NVIDIA GTX650 graphics card. It runs on a Ubuntu 12.04 (32 bit) and uses CUDA-5.5. The time was measured by means of the 'time.h' library (Unix/POSIX) and is the elapsed real time. It was averaged over 15 measurements for each event. The results are presented in Figure 5.8 for a number of 12000 processed ARGONTUBE events. The graph on top shows the time measurements for the CPU and the GPU. The ordinate axes have scales differing by a factor of 50. The histogram on the bottom illustrates the distribution of the number of pixels that were above threshold and subject to the Hough transform for real ARGONTUBE events. Values for the mean (most probable) number of pixels are about 4800 (2100). The total number of pixels for one ARGONTUBE event is given by $64 \cdot 8192 = 524288$. The time measurements corresponding to the CPU code are composed only of the computation of

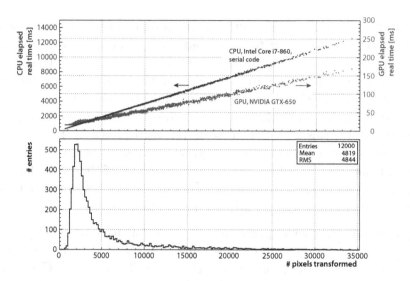

Figure 5.8: *Top:* Measured CPU and GPU computation times (real elapsed time) for the PCLines Hough transform as a function of the number of pixels that were transformed. Note the factor 50 difference in the two time scales. *Bottom:* Distribution of the number of transformed pixels for the processed ARGONTUBE events.

the bare Hough transform and the filling of the accumulator array, while for the GPU the time needed for the data transfer from host to device and vice versa (memory overhead) is included. As expected, the computation time for the serial code increases linearly with the number of pixels that must be transformed. The same is true for the CUDA code. However, in the latter case, the linearity cannot a priori be expected due to the memory overhead which could depend on the amount of data transferred from device to host and vice versa, and thus on the number of pixels that are transformed. A powerful tool for the analysis and optimization of CUDA code is provided by NVIDIA and is called the Visual Profiler. Amongst others, this software is able to separate the time taken for the individual tasks, such as copying input data from host to device, kernel execution time and copying the output data back to the host. The memory overhead in the present example was measured to vary from a negligible $1\,\mu s$ to $100\,\mu s$ for the transfer from host to device (input data) and roughly $10\,ms$ from device to host (output data, i.e. accumulator array). The size of the input data and thus the time needed for the transfer depends on the number of pixels that are to be transformed. Concerning the output data, this is not the case as its size is given by the number of entries in the accumulator array, which remains constant. It was found that the time difference between the sum of the values for the memory overhead and the data shown in Figure 5.8 can be almost fully attributed to the actual kernel execution time.

For a small number of pixels, one expects that it is not beneficial to copy forth and back data to device memory to perform only relatively few arithmetic operations on the GPU (see Section 5.2). Certainly, this is the case when dealing with a large amount of input or output data as the computation time may become negligible and the memory overhead strongly dominant. This does not occur for the PCLines Hough transform algorithm. Figure 5.9 shows the ratio of the computation times t_{CPU}/t_{GPU} and is a measure for the speed-up that is achieved as a function of the number of transformed pixels. Even for a small number of pixels (< 1000), the GPU processes the data faster than the CPU. However, the speed-up is below a factor 20, since the hardware resources of the graphics card are not fully exploited and the memory overhead is relatively large (more than $50\,\%$) at these computing times. The discontinuities in the plot occur every 2048 pixels and originate from the CUDA/GPU thread management. A discussion is out of the scope of this thesis and the reader is referred to [98, 99].

Straight lines were fitted to the data points in Figure 5.8 to derive the speed-up which is approached with an increasing number of transformed pixels. In the limiting case, the GPU is faster by a factor of

$$S = 79.7 \pm 0.1 \tag{5.2}$$

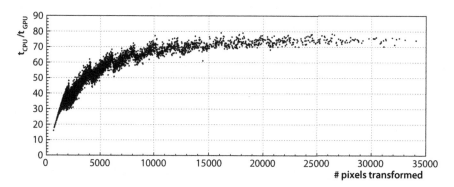

Figure 5.9: The speed-up of GPU versus CPU (serial) code as a function
of the number of transformed pixels. The discontinuities
occur every 2048 pixels and originate from the CUDA/GPU
thread management.

compared to serial code executed on the CPU. At the mean number
of roughly 4800 pixels for ARGONTUBE events, the speed-up is $54.0 \pm$
0.5 and at the most probable value of about 2100 pixels, it reaches
36.6 ± 0.5.

The PCLines algorithm is already optimized in terms of computing
efficiency as it relies only on simple arithmetic operations and fully
avoids time-consuming mathematical functions (e.g. trigonometric).
Moreover, the graphics card that was used is a low-budget device.
The latest high-end models have much larger hardware resources
and are particularly designed for GPGPU computing for scientific
applications. Finally, the code can be optimized which would result
in an additional speed-up. In conclusion, the advantages of using a
GPU for computational problems with a high arithmetic intensity are
obvious. The task of track reconstruction, e.g. for events recorded
with a large LArTPC, can strongly benefit from the use of GPUs.
Even though for ARGONTUBE events the speed-up is far from the
hardware limit, due to the relatively small number of transformed
pixels, it is still significant.

6 Study of cosmic muon events

A large number of cosmic ray events have been acquired during the three ARGONTUBE runs in April, June and October 2013 with different read-out configurations. While in April the read-out was realized with only 'warm' preamplifiers, half the number of channels were equipped with LARASIC4 cryogenic electronics for the run in June to study their performance in a real LArTPC. The LARASIC4, developed by BNL [56, 62–64], demonstrated outstanding performance and yielded high-quality data. Thus, for the latest run carried out in October, the ARGONTUBE read-out was fully upgraded to cryogenic preamplifiers. The data recorded in June and October were evaluated to determine the muon energy deposition per unit track length dE/dx in liquid argon, both for the 'warm' and the cryogenic read-outs. The idea was to verify the test bench measurements of the transimpedance for the LARASIC4, a quantity needed for the charge to energy conversion, and to take a first look at the calorimetric capabilities of ARGONTUBE, which may become relevant for future studies of muon events. These measurements were also suited to test the procedure used to determine the longitudinal electric field introduced in Chapter 4 and the track finder described in Chapter 5. The results obtained for dE/dx are discussed in Section 6.1. To explore the possibility of estimating the momentum of muons producing only partially contained ionization tracks in the detector, the procedure described in [9] was implemented and applied to ARGONTUBE events. The knowledge of the muon momentum distribution could be used to perform a detector self-calibration [101] to cross check the transimpedance measurements of the preamplifiers. An overview on the method of momentum estimation by multiple Coulomb scattering is presented in Section 6.2.

6.1 Linear energy transfer of muons in liquid argon

Cosmic muon events recorded in June and October 2013 were considered for the study of the energy loss per unit length dE/dx in liquid argon. The first stage of the analysis was done by means of the reconstruction software described in Chapter 5. The output are track candidates found from the collection wireplane data (x-z projection). To measure the energy deposition per unit length, the length of the ionization track per wire, or equivalently, the track inclination in 3D, must be known. It cannot be determined from the collection plane data alone and the induction view (y-z projection) must be considered as well. The two projections are combined by mapping the hits found respectively in the collection and induction views by means of their drift coordinates. The identification of induction hits and the extraction of information from the induction bipolar signals is much more difficult than for collection plane data. To transform the bipolar into unipolar pulses simplifying the search for hits and the determination of their positions, the induction waveforms were convoluted with an average induction signal, generated from muon tracks nearly transverse in the y-z projection. The convolution method is an approximative approach and works best for transverse tracks, but also yields useful results otherwise. If one was interested in obtaining not only spatial, but also calorimetric information from the hits of the induction view, a more sophisticated procedure would be needed, based on a de-convolution of the induction plane waveforms with a carefully determined δ-response function for the read-out [51]. For each of the track candidates found in the collection plane, the induction view was searched for a corresponding track and a match was found if the beginning and the endpoints of the tracks, described by their drift coordinates, were in agreement. Figure 6.1 shows the collection (top) and induction (bottom, after convolution) views of a typical track used for the of dE/dx study. The drift time has already been translated to the spatial coordinate z by means of the longitudinal electric field

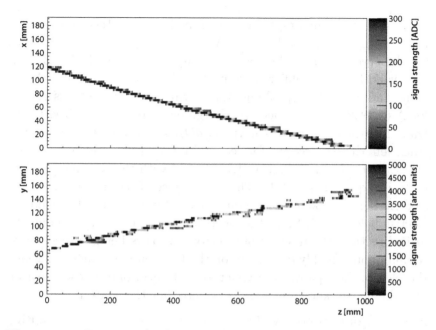

Figure 6.1: Collection (top) and induction (bottom, after convolution) views of a track used for the measurement of dE/dx.

model described in Chapter 4. The hits found in the two projections are fitted with straight lines (see Fig. 6.2) which allows the calculation of the track direction and position in 3D. From these measurements, and taking into account the wire spacing in ARGONTUBE, the track length per wire can be determined.

To obtain calorimetric information, the hits on each wire of the collection plane were looked at individually. Since the tracks mainly have strong inclinations, the preamplifier input signals are far from δ pulses. Thus, the preamplifiers operate in the current-sensitive mode and the charge of a pulse must be determined from its integral. After correcting for the finite charge lifetime, known from purity measurements with the UV laser (see Chapter 7), the integral values are transformed into the equivalent charge amounts ΔQ (collected charge per wire) by means of the preamplifier transimpedance Z. Test bench measurements gave $Z_{cryo} = 117 \pm 3 \, \text{mV/nA}$ and $Z_{warm} = 13 \pm 1 \, \text{mV/nA}$ for the cryogenic and the warm preamplifiers respectively. Z_{cryo} is only valid for an amplifier configuration with $G = 25 \, \text{mV/fC}$ (charge gain) and $T_p = 3 \, \mu s$ (peaking time), which are the settings chosen for ARGONTUBE. The collected charge per wire is normalized to the track length per wire which yields dQ/dx, with dx being the unit track length. To account for recombination losses and to translate dQ/dx into dE/dx, Equation 1.12 is used with the parameter values given in Equation 1.10 [31]. The dependence of recombination on the electric field is taken into account again by use of the longitudinal electric field model. Since the electric field varies strongly along the drift direction, the fraction of electron-ion pairs being subject to recombination is highly dependent on the location in the detector where the electron-ion pairs were generated. Hence, only tracks with an unambiguously known t_0 were considered for the analysis.

A number of respectively 758 and 381 tracks were selected resulting in 12303 and 3731 pulses that were analyzed for the cryogenic and the 'warm' read-outs. The mean numbers of pulses per track are thus roughly 16 and 10 respectively. The discrepancy does not originate from a different selection or angular distribution of the tracks. The data used for the analysis of the 'warm' read-out were taken in June when 10 channels out of 64 of the wireplane in collection mode were equipped with cryogenic electronics and thus could not be taken into account. The resulting distributions for dE/dx are shown in Figure 6.3. They indicate an average energy loss per unit length for muons in liquid argon of

$$\langle dE/dx \rangle_{cryo} = (1.82 \pm 0.01 \, \text{(stat.)} \pm 0.34 \, \text{(sys.)}) \, \text{MeV/cm}, \quad (6.1)$$

$$\langle dE/dx \rangle_{warm} = (2.23 \pm 0.02 \, \text{(stat.)} \pm 0.42 \, \text{(sys.)}) \, \text{MeV/cm}, \quad (6.2)$$

in agreement with literature values (see Tab. 1.1). As suggested by [51], a fit with a Landau function convoluted with a Gaussian was performed to find the most probable dE/dx for muons in liquid argon.

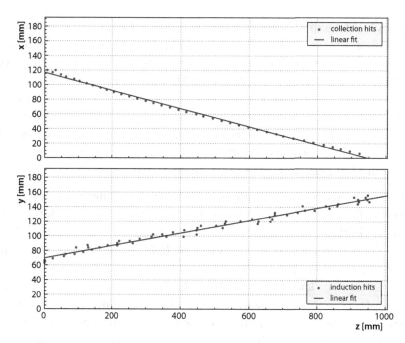

Figure 6.2: Hits identified in the collection (top) and induction (bottom) views for the example track in Figure 6.1. They are fitted with straight lines to determine the direction, position and length (here about 960 mm) of the track within the sensitive detector volume.

While the Landau function describes the energy loss of the particles in the medium [21], the Gaussian accounts for effects originating amongst others from electronics noise. The results are

$$dE/dx^{m.p.}_{\mathrm{cryo}} = (1.44 \pm 0.26 \,(\text{sys.})) \text{ MeV/cm}, \qquad (6.3)$$

$$dE/dx^{m.p.}_{\mathrm{warm}} = (1.56 \pm 0.27 \,(\text{sys.})) \text{ MeV/cm}, \qquad (6.4)$$

with negligible statistical errors and in agreement with each other. These values depend on the momentum distribution of the muon sample that was studied. Reference [101] shows the relationship between $dE/dx^{m.p.}$ and the momentum for muons in the range of zero to 10 GeV for liquid argon as obtained from a Monte Carlo simulation. Values of about 1.6 MeV/cm to 1.75 MeV/cm are given for muon momenta > 0.4 GeV including a dependence on the track inclination. A value of $dE/dx^{m.p.} = 1.82 \pm 0.10$ MeV/cm was reported by [102] and was obtained from cosmic muon events. δ-rays, which coincide with the main ionization track (low energy) or move away from it (high energy, few MeV) were not subtracted. It has been shown by [51] that taking into account δ-rays does not change the parameters of the Landau-Gaussian fit within statistical uncertainties.

The systematic errors include contributions from uncertainties in the electric field model, the track inclination, the transimpedance, the charge lifetime, the parameters in Equation 1.10 and in the value W_i. The uncertainty in the electric field model has a result on both the fraction of recombination and on the position resolution of the detector in the z coordinate. The latter causes an uncertainty in the track length per wire. While the contributions from the fit parameters of the recombination box model and from the charge conversion factor are almost negligible, the major uncertainty in dE/dx originates from the accuracy in the track length per wire and from the electric field model. For future studies, the systematics must be reduced. The first improvement to be made is the processing of the induction plane hits. The approach used here is much simplified. Even though the conversion of bipolar to unipolar pulses can be done by a convolution, it is an approximative method which results in a loss of spatial information. Moreover, the determination of the track inclination does not use full 3D tracking but relies on the straight line fits to the collection and induction hits. By using the method suggested in [51], which needs a thorough calibration of the read-out, more precise information about the spatial coordinates of the induction hits can be extracted resulting in a much higher accuracy in the track length per wire and thus in reduced systematics for dE/dx. With a correct induction signal processing, it would even be possible to obtain calorimetric information from the induction hits and to combine these with the collection hits to improve the energy resolution.

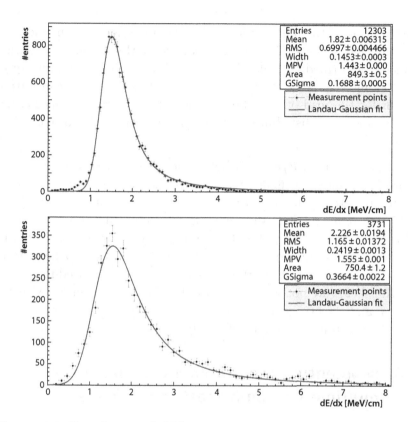

Figure 6.3: Distribution of dE/dx obtained with the cryogenic (top) and the 'warm' (bottom) read-outs for a number of 758 and 381 tracks respectively. A Landau function convoluted with a Gaussian is fitted to the data to determine the most probable value for dE/dx (MP).

The mean values given in Equations 6.1 and 6.2 were found using the test bench measurements of the amplifier transimpedances. To do a detector self-calibration [101, 103], where the amplifier transimpedance is obtained from detector events (e.g. the muon sample), the measured most probable value for the collected charge per unit length $dQ/dx_{\mathrm{meas}}^{m.p.}$ can be used. In liquid argon, the most probable value for the energy deposition per unit length $dE/dx^{m.p.}$ depends on the momentum distribution of the muons selected for the analysis. If the momentum distribution of the studied tracks is known, a Monte Carlo simulation can be performed to yield the expected $dE/dx_{MC}^{m.p.}$ in liquid argon. By taking the ratio of $dQ/dx_{meas}^{m.p.}$ and $dE/dx_{MC}^{m.p.}$, the calibration factor for the read-out can be obtained. Possible ways of measuring the momentum distribution of the muon sample are discussed in Section 6.2.

By comparing the dE/dx distributions found for the cryogenic and the 'warm' read-outs, one finds that the latter has a larger width and more events towards higher dE/dx. The widths of the Gaussians indicate 0.17 MeV/cm and 0.37 MeV/cm for the cryogenic and the 'warm' electronics respectively. Partly, the differences can be attributed to the electronics noise being different in the two cases. Other effects most likely originate from the track reconstruction procedure. If the hit thresholds are not set equivalently for the events acquired with the 'warm' and the cryogenic read-outs respectively, δ-rays (in particular) may be taken into account differently.

6.2 Momentum measurement by analysis of multiple Coulomb scattering effects

Often, particles are not stopped within the sensitive volume of a detector and their ionization tracks are thus only partially contained. As a result, obtaining kinematic information is not possible by integrating the total amount of charge deposited along the path. For event reconstruction and particle identification, it is essential to have calorimetric information and to know the momenta of the interacting particles. There is a possibility to measure the momentum of a through-going or exiting particle by analyzing the many small angle scatterings (multiple Coulomb scattering) that it undergoes when traversing the sensitive detector volume. Reference [9] studies and explains in much detail two different methods which allow the measurement of the momentum on a track by track basis, both with Monte Carlo and real LArTPC data. The first one is a so called classical approach. An ionization track is split into segments and the RMS width θ_0 of the angular distribution of the plane-projected scattering angle, obtained from the segments along the track, is determined as a function of the segment length s. The relation between θ_0, s and the product of momentum and velocity of the traversing particle is given by Equation 1.5. The classical approach measures the momentum with an uncertainty of about 25 % in the range 0.5 GeV to 3.0 GeV [9]. A particle that passes through a medium constantly loses energy and momentum along its path. The classical approach does not compensate for these losses and the initial momentum is thus underestimated. There is, however, the possibility to measure the total energy loss of the particle in the detector by integrating the signal along the track, and to apply an a posteriori correction. The second method uses a more sophisticated approach based on Kalman filtering and achieves a resolution of about 10 % in the same range of momenta. The more than twice better resolution originates from the Kalman filter being able to account for detector noise and spatial resolution as well as for physics processes like energy loss and multiple Coulomb scattering (MCS) along the particle track. It decouples the detector noise and MCS effects in the best possible manner. For both methods, the uncertainty in the measured momentum increases for shorter tracks and for momenta below 0.5 GeV, which is mainly due to reduced statistics. However, these particles produce tracks shorter than 1 m in liquid argon and are usually fully contained in large LArTPCs. Their energies and momenta can be obtained from their range and energy deposition. Another issue is observed for very high momenta of the order of 10 GeV or more. For such highly energetic muons, the MCS effects are small and they can no more be disentangled from the detector noise.

The procedure used to derive the momentum for muons recorded with ARGONTUBE is given by the classical approach and consists of the following steps [9]

1. The (through-going) muon track is reconstructed.

2. δ-ray hits are identified and removed.

3. The track is split into segments of a certain length s.

4. The angular direction θ_i of each segment i is determined by a straight line fit.

5. The difference between the directions of each consecutive pair j of segments is determined to yield an angle $\Delta\theta_j$.

6. The RMS of the distribution of the $\Delta\theta_j$ is computed after cutting away 2 % of the distribution tails, since only the central 98 % follow a Gaussian (Molière theory).

7. $\theta_0^{\mathrm{meas}}(s)$ for the given segment length s is then determined as the RMS of the $\Delta\theta_j$ that remain after removal of the values that are more than 2.5 times the RMS away from the mean (removal of outliers).

Equation 1.5 describes the dependence of the RMS width θ_0^{MCS} of the angular distribution for MCS as a function of the segment length s and of the product of particle momentum and velocity βcp. However, it cannot be directly applied to $\theta_0^{\mathrm{meas}}(s)$ as there is a detector noise contribution θ^{noise} in addition to the MCS effects, originating from the limited spatial resolution. For short segment lengths, the MCS effects are small and θ^{noise} dominates. The noise contribution depends only on the segment length and is given by [9]

$$\theta^{\mathrm{noise}} = C_{\mathrm{noise}} \cdot s^{-3/2}, \tag{6.5}$$

where C_{noise} is a constant of proportionality depending on the detector tracking capabilities. The measured RMS width of the angular distribution is composed of the noise and MCS terms

$$\theta_0^{\mathrm{meas}} = \sqrt{\left(\theta_0^{\mathrm{MCS}}\right)^2 + \left(\theta^{\mathrm{noise}}\right)^2}. \tag{6.6}$$

The model in Equation 6.6 can be used to fit the measurement points $\theta_0^{\mathrm{meas}}(s)$ as a function of the segment length. It has only two free parameters C_{noise} and βcp. If the particle producing the ionization track is known, its momentum can be calculated by means of its mass.

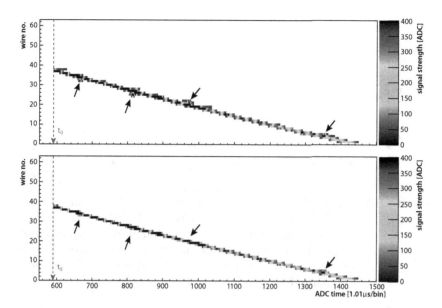

Figure 6.4: *Top:* Reconstructed track in the collection plane. It is dressed with low and high (arrows) energy δ-rays. *Bottom:* The same track after removal of high energy δ-rays.

The example track depicted in Figure 6.4 is used to illustrate the procedure. The top view shows the reconstructed track as it results from the track finder (see Chapter 5). The dashed line indicates the t_0 of the event and corresponds to the location of the wireplane. δ-rays appear along the track. While low energy δ-rays produce hits coinciding with the main ionization track, those of higher energies (a few MeV) have a much higher range and can produce an ionization track well separated from the one generated by the main particle (see arrows). Particularly for short segment lengths, δ-rays would strongly bias the angular direction of a segment and hence have a strong impact on the momentum estimation. They must be removed. For each wire, the signal waveform is scanned for the number of hits. A hit is given by a pulse (local maximum) which has a certain width in time, defined by the limits where the signal strength at the low and

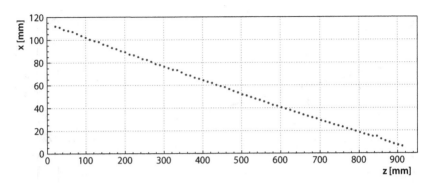

Figure 6.5: The track given in Figure 6.4 on the bottom is sliced along
the drift time axis. The drift time is translated to the spatial coordinate z by means of the longitudinal electric field
model. For each slice (width of 10.1 μs), the center of gravity
in the x coordinate is determined and plotted. Error bars
are omitted for reasons of visibility.

high ends of the pulse falls below a chosen threshold. The choice of
the threshold value is critical. If it is set too low, local maxima are not
well separated, while for high thresholds, information about the pulses
is lost which results in worse spatial resolution. A threshold of 10 % of
the local maximum amplitude was found to be a good compromise in
the present case. On each wire, the hits not coinciding with the main
track are removed. For the example track in Figure 6.4, the result
after δ-ray removal is shown in the bottom plot. Hits originating
from high energy δ-rays are well separated from the main track and
are thus reliably removed. Nevertheless, some δ-ray pulses remain,
as they overlap with the main track. They are taken into account
partly by the detector noise term in Equation 6.6 and in step 7 of the
momentum estimation procedure.

The track in Figure 6.4 on the bottom is broken up into slices of
width 10.1 μs along the time axis and the drift times are transformed
to spatial coordinates z by means of the longitudinal electric field
model (see Chapter 4). The wire number is translated into the spatial
coordinate x using the ARGONTUBE wire spacing of 3 mm. Within

each slice, the center of gravity in x is determined and plotted versus z. The resulting graph is shown in Figure 6.5. In this example, the particle has left a track of about 960 mm length in the ARGONTUBE sensitive volume, which is just enough for the momentum estimation method to work with a reasonable accuracy [9].

To find meaningful results for the momentum, the set of segment lengths must be chosen appropriately. For the example track, values reaching from a minimum length of 25 mm to a maximum of 165 mm with a spacing of 20 mm were used. The lower limit is determined by the density of measurement points along the track in Figure 6.5. The angular distribution at this length is governed by detector noise and there is no reason in making even shorter segments. The maximum value is dependent on the track length and it should be chosen such that there are enough entries in the angular distribution for determining the RMS value.

For each value of the segment length, the data points in Figure 6.5 are grouped accordingly (step 3). The points in each segment are fitted with a straight line (step 4) and for consecutive segments, the differences in the direction are accumulated in an array and the RMS value is calculated after removing 2 % of the distribution tails and outliers (steps 6 and 7). The obtained values for θ_0^{meas} as a function of the segment length are shown in Figure 6.6. The error bars include the propagated uncertainties in the detector spatial resolution. By fitting the model given in Equation 6.6 to the data, one finds $C_{\text{noise}} = 0.08 \pm 0.05\,\text{cm}^{3/2}\,\text{rad}$ and $\beta cp = 2.46 \pm 0.94\,\text{GeV}$. Using the muon mass $m_\mu = 105.7\,\text{MeV}$ [21], one finds that $\beta = 0.9991$ and thus $p \approx \beta cp$ ($c = 1$).

To illustrate the effect on θ_0^{meas} for a particle with a different momentum, a second track is shown in Figure 6.7. Compared to the previous example, MCS effects are more pronounced, especially for longer segment lengths. Consequently, the muon had a lower momentum than the one shown first in Figure 6.4. This is confirmed by the plot in Figure 6.8. The fit results are $C_{\text{noise}} = 0.10 \pm 0.05\,\text{cm}^{3/2}\,\text{rad}$ and $\beta pc = 0.96 \pm 0.09\,\text{GeV}$. Taking into account the muon mass,

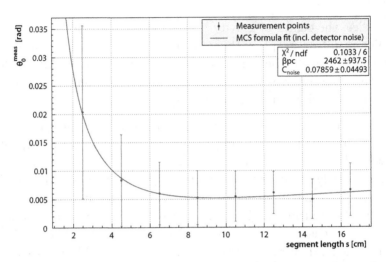

Figure 6.6: θ_0^{meas} as a function of the segment length s for the track in Figure 6.4. For short segments, the detector noise starts to dominate while for longer segment lengths MCS effects become visible. The model in Equation 6.6 is fitted to the data.

$p = 0.95 \pm 0.09\,\mathrm{GeV}$. The constants of proportionality C_{noise} describing the detector noise are in agreement for the two tracks.

The method described here has been proven to work in [9] by means of a Monte Carlo simulation. However, a validation of the implementation used for ARGONTUBE is still pending. A first possibility to verify the method would be to process the cosmic muon tracks that have been collected so far to obtain angular, track length and momentum distributions. By means of a Monte Carlo simulation [104] that uses the same distributions as well as the same cuts for the analysis, the method could be verified. An alternative way for validation would be to use the muons stopping in the ARGONTUBE. Their energy is known from their range and energy deposition and could be compared to the values obtained from the MCS method. The second idea was applied to ICARUS data by [9] to verify the more sophisticated Kalman filtering approach for momentum estimation on a track by track basis.

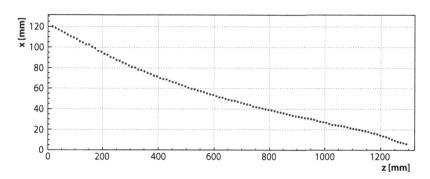

Figure 6.7: Reconstructed collection plane hits after δ-ray removal for a track with a lower momentum than that shown in Figure 6.4.

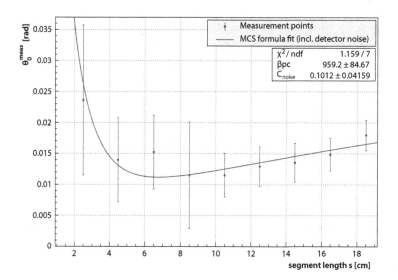

Figure 6.8: θ_0^{meas} as a function of the segment length s for the track in Figure 6.7. Compared to Figure 6.6, the RMS width of the angular distribution is larger and has a steeper increase towards longer segments, meaning that MCS effects are more pronounced. The track thus originates from a muon with lower momentum.

7 UV laser methods and measurements

The ARGONTUBE UV laser system is a versatile instrument not only to infer information about the electric field, but also to measure the level of impurities present in the system. A good knowledge of the latter is required for the reconstruction of calorimetric information as the attenuation by electron attachment must be compensated for. One method, which is explained in [24, 25], is based on measuring the signal attenuation as a function of the drift time for UV laser induced ionization tracks. A different approach was chosen here, based on the principles of liquid argon purity monitors [52, 54]. It is discussed in Section 7.1. The obtained result is compared to the value found from the analysis of cosmic muon tracks. Other parameters which are accessible by means of UV laser induced events are the transverse [8] and longitudinal diffusion coefficients in liquid argon as a function of the electric field strength. Diffusion processes, both transverse and longitudinal, are important aspects to consider when building and operating large LArTPCs as they have an impact on the spatial resolution of the device. A measurement of the longitudinal diffusion coefficient D_L is presented in Section 7.2 for electric field strengths reaching from 120 ± 49 (RMS) V/cm to 210 ± 84 (RMS) V/cm.

7.1 Liquid argon purity measurements

When shooting a UV laser pulse on the TPC cathode, a certain amount of charge Q_0 is released from the metal by means of the photoelectric effect. In the presence of the electric field, this charge cloud drifts across the full length of the TPC and is eventually collected at the wireplane. A certain number of electrons, initially corresponding to the cloud, are attached by impurities during the drifting process and are thus lost resulting in a collected charge amount of $Q_0 e^{-t_d/\tau} \leq Q_0$. t_d denotes the drift time and τ is the characteristic time constant describing the concentration of oxygen equivalent impurities in liquid argon in the sensitive volume (often called 'charge lifetime' in this context). Concerning liquid argon purity monitors [52, 54], Q_0 is usually obtained by means of a wire grid situated in proximity of the cathode. The charge cloud passing through the grid induces a bipolar signal of a strength depending on the amount of charge. t_d is given by the time difference between the passage through the grid near the cathode and the collection of the charge at the anode. Having the knowledge of these values, τ can be calculated.

For ARGONTUBE, Q_0 is not measured and a different approach must be chosen instead. The idea is to generate at different electric field strengths E_i the same amount of charge Q_0 and to consider the amount of collected charge Q_i as a function of the drift time $t_i = t_i(E_i)$. The latter relation is a consequence of the electric field dependent electron drift speed (see Section 1.6). Assuming that $Q_0 = \text{const.} \forall i$ and attenuation of the charge cloud is due to electron attachment only, one expects that the measurement data (Q_i, t_i) obey the model (see Eq. 1.25)

$$Q(t_i) = Q_0 \cdot e^{-t_i/\tau}, \qquad (7.1)$$

with τ and Q_0 being the free parameters. In principle, measuring at two different E_i is enough to calculate Q_0 and τ. However, it is recommended to take a larger number of measurement points and to obtain Q_0 and τ from the best fit of Equation 7.1 to the data.

There are certain difficulties that have to be addressed when using this approach. First of all, the amount of charge released from the cathode by means of the UV laser is not strictly a constant, but is subject to fluctuations. Some of these can be attributed to the laser, either to an unstable light source (see Section 2.4) or to the behaviour of the beam in liquid argon (focusing and defocusing effects). By recording

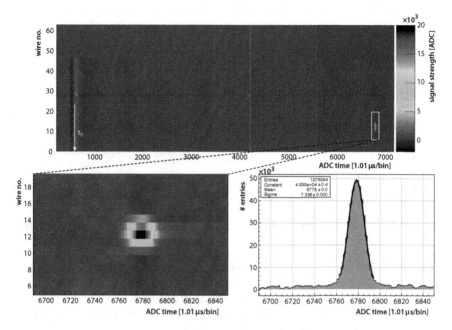

Figure 7.1: *Top:* Sum of 200 UV laser shots on the gold-plated ARGON-
TUBE cathode. t_0 is indicated on the left, while the charge
cloud released from the cathode is seen on the right (box).
Bottom left: Zoomed view of the endpoint. *Bottom right:*
Gaussian fit on the sum of wires 10 to 15.

a significant number of single laser shots at the same conditions, it
is possible to obtain a distribution of collected charge which can
be fitted with a Gaussian. Secondly, concerning ARGONTUBE, the
high voltage is not delivered by a power supply but by a Greinacher
circuit (see Section 2.2) and there is no neat way of lowering the drift
field since the circuit cannot be discharged externally. Instead, one
must await a spontaneous discharge caused by an accidental dielectric
breakdown from the TPC field cage to the cryostat wall. Starting
from the low voltages, laser events can be taken at different electric
field strengths by charging the high voltage circuit step by step to its
original value. Thirdly, even though the charging frequency of the

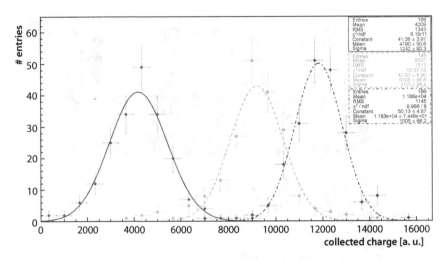

Figure 7.2: Histograms of collected charge for three different drift times of about 8.73 ms (solid), 7.12 ms (dashed) and 6.59 ms (dot-dashed) for a constant initial charge Q_0. The strong dependence of collected charge on the drift time results from electron attachment by impurities.

Greinacher circuit has been increased from 50 Hz to 500 Hz, it takes roughly 7 minutes to reach a charged state for a given voltage setpoint and one purity measurement series takes about 1 h to be completed. During this period, the charge released from the cathode Q_0 must be held constant. An accidental change in the laser beam power or in the alignment (due to vibrations) can already have severe consequences as this typically leads to a change in Q_0 and makes the measurement series (at least partially) useless. These effects were in fact observed. One way to deal with them is to perform a cross check measurement at the end of a series.

To obtain one charge lifetime measurement, sets of 200 single UV laser shots at seven or eight different electric field strengths were acquired and analyzed. Figure 7.1 shows the sum of 200 single laser shots that were recorded under the same conditions. Two relatively strong signals are visible. The first one is located at $t_0 = 832\,\mu s$ (bin 590) and corresponds to the location of the wireplane. The second signal of interest is at $t_c = 9557\,\mu s$ (bin 6778) and results from the charge cloud (also referred to as the endpoint) generated at the cathode. Figure 7.1 on the bottom left shows a zoomed view. The drift time across the full length of the chamber at the given electric field is $t_c - t_0 = 8725\,\mu s$. A precise measurement of these values is obtained by fitting the endpoint signal with a Gaussian (see Fig. 7.1, bottom right). As opposed to the events generated to estimate the transverse component of the electric field, where it was essential to know the path of the laser beam (compare Fig. 4.2), the laser power had been strongly attenuated in the present case so that there was practically no ionization track along the beam path. This allows much more reliable decoupling of the signal generated by argon ionization in proximity of the cathode from the actual charge cloud induced by the photoelectric effect. This simplifies the analysis of the charge lifetime and particularly that of the longitudinal diffusion coefficient.

Apart from the drift time t_i, one is interested in the collected charge for every single laser shot. A quantity which is proportional to the collected charge is given by the integral of the recorded signal pulse. To set the integration boundaries consistently for all the measurements, the width σ_i of the Gaussian fit on the sum of the pulses was used (see Fig. 7.1, bottom right). The individual pulses were integrated in the range $[t_{c,i} - 3\sigma_i,\ t_{c,i} + 3\sigma_i]$ and the results were accumulated in a histogram. Figure 7.2 depicts histograms which represent the distributions of collected charge along with the Gaussian fits for three different drift field strengths. The only parameter that was changed for these measurements was the electric field strength and therewith the drift time. Q_0 was held constant. The collected charge shows an exponential dependence on the drift time which originates from the attachment of electrons by impurities in the sensitive detector volume. This becomes evident when plotting the mean values Q_i from the Gaussian fits versus the corresponding drift times t_i for all the measurements of one series (see Fig. 7.3). The error bars of the individual measurement points are only of a statistical nature. They

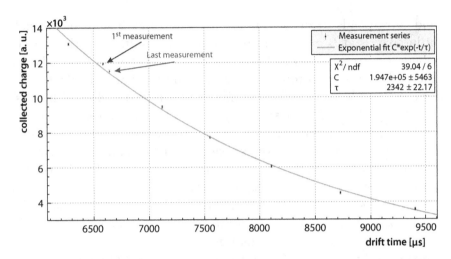

Figure 7.3: One example of a charge lifetime measurement with the UV
laser. The collected charge is plotted versus the drift time.
Data points are only attributed statistical errors. The solid
line shows the exponential fit (see Eq. 7.1). The two cross
check measurements (first and last) of the series allow to
verify self-consistency of the data.

result from the Gaussian fits to the collected charge distributions. The
first and the last data points of the series are cross check measurements
and were separated by a time of about 1 h. They were performed at
similar electric field strengths and thus at similar drift times. Since
they agree on the amount of collected charge, the measurement series is
consistent. During data taking, the laser alignment was stable and the
level of impurities did not change, given the accuracy of the method.
The charge lifetime obtained from the best fit is $\tau_{\text{Laser}} = 2.34 \pm 0.03$ ms,
corresponding to an oxygen equivalent impurity concentration of about
0.13 ppb (see Eq. 1.26).

To study the plausibility of the results obtained with the newly
introduced method, the argon purity was also estimated from a sample
of straight cosmic muons that were recorded in a time frame of
18 h, right after the measurement series presented in Figure 7.3. An

approach similar to the one described by [50, 51] was used. Muon events reconstructed by means of the track finder (see Chapter 5) and located in the upper drift region (0 ms to 1 ms) were scanned for hits in the same manner as discussed in Section 6.1. The tracks were selected for an unambiguously known t_0 and available induction view to determine their inclination. The restriction to the upper detector region is due to a nearly constant longitudinal electric field E_L that shows a drop of at most 5 % in this range (compare Figure 4.3). This legitimates not correcting for the field dependent recombination effects. The scatter plot in Figure 7.4 illustrates the collected charge (hit integrals) corrected for track inclination as a function of the drift time for the 1147 tracks selected from 6000 recorded events. It consists of 17545 individual pulses that were analyzed. A decrease with increasing drift time is clearly visible and can be fully attributed to the attachment of the drifting electrons to impurities present in the sensitive volume of ARGONTUBE.

To quantify the dependence of collected charge on the drift time, the 2D scatter plot was divided into 32 time slices each having a width of 30.3 μs. The width is chosen as small as possible to make the dependence of collected charge on the drift time negligible within the slice, but such that it contains enough data to fit them reliably. For each slice, the most probable value of collected charge and its error, both obtained from a Landau-Gaussian fit, is shown in Figure 7.5. An exponential fit to the 32 measurement points yields a charge lifetime of $\tau_{Cosmics} = 1.91 \pm 0.05$ ms. The given error is purely statistical and results from the exponential fit.

The ARGONTUBE recirculation system was switched off during the time period of 18 h cosmic data taking meaning that the amount of impurities had been steadily increasing. The value $\tau_{Cosmics}$ thus marks a mean charge lifetime over the given time frame. It is to be compared to $\tau_{Laser} = 2.34 \pm 0.03$ ms obtained from UV laser data acquired before. The discrepancy is explicable by the recirculation system not being in operation, combined with the large amount of time needed to record enough muon events to get a reliable estimate

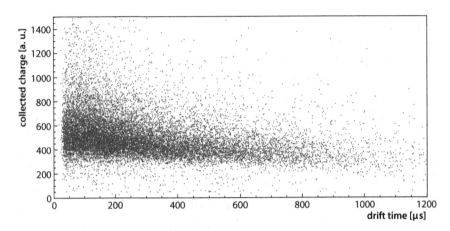

Figure 7.4: Scatter plot of the hit integrals in the upper region of the detector as a function of the drift time. The decrease in the collected charge with increasing drift time originates from electron attachment by impurities present in the sensitive volume.

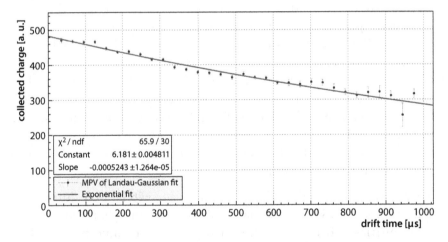

Figure 7.5: Plot of the most probable values obtained from a Landau-Gaussian fit [51] to the distribution of collected charge in each time slice in Figure 7.4. The exponential fit indicates a charge lifetime of $\tau = 1.91 \pm 0.05\,\text{ms}$.

of the charge lifetime. An additional measurement of τ with the UV laser right after cosmics data taking yielded $1.24 \pm 0.02\,\text{ms}$, consistent with a constantly increasing level of impurities. Another difference between the two measurement methods is that the muons are only considered in the uppermost region of the drift volume, while with the laser induced charge cloud the concentration of impurities along the full drift distance is taken into account. Even though it has not yet been confirmed, the concentration of contaminants is expected to be higher in proximity of the anode, where the largest amount of impurities originates from (gas phase).

7.2 Measurement of the longitudinal diffusion coefficient

Relevant theoretical contributions in modeling the diffusion processes of electrons in liquid argon originate from Cohen and Lekner [105], and Atrazhev and Timoshkin [106]. The theory indicates values for the transverse and longitudinal diffusion coefficients of $D_T = 16.3\,\text{cm}^2/\text{s}$ and $D_L = 6.2\,\text{cm}^2/\text{s}$ respectively at an electric field strength of $500\,\text{V/cm}$. There are only a few experimental studies on this subject. One reason is that diffusion is low in noble liquids compared to gases and hence large drift times (and drift distances) are necessary to obtain reliable measurements. Reference [107] reports values for the mean electron agitation energy (see Eq. 1.20) for electric fields in a range from $2\,\text{kV/cm}$ to $10\,\text{kV/cm}$. The values can be translated into $D_T = 3\,\text{cm}^2/\text{s}$ at $1\,\text{kV/cm}$ to $D_T = 16\,\text{cm}^2/\text{s}$ at $10\,\text{kV/cm}$ [20]. Reference [108] reports slightly higher values, namely $D_T = 13\,\text{cm}^2/\text{s}$ at $1\,\text{kV/cm}$ and $D_T = 9\,\text{cm}^2/\text{s}$ at $300\,\text{V/cm}$. Another measurement was done with the ARGONTUBE detector with UV laser ionization tracks and gave $D_T = 4.2 \pm 0.4\,\text{cm}^2/\text{s}$ at a field strength of about $200\,\text{V/cm}$ [8]. Concerning the longitudinal diffusion coefficient, one value has been published by the ICARUS collaboration who obtained $4.8 \pm 0.2\,\text{cm}^2/\text{s}$ from muon tracks, and averaged over electric field strengths from $100\,\text{V/cm}$ to $350\,\text{V/cm}$ [5].

The large number of UV laser events that were acquired at different electric field strengths to determine the concentration of impurities in liquid argon can be used to measure the longitudinal diffusion coefficient and its dependence on the electric field strength. A laser pulse has a duration of 4 ns to 6 ns only (see Sec. 2.4) and the charge cloud released from the cathode by means of the photoelectric effect can be considered to initially follow a δ-distribution in the longitudinal direction. During the drift, the cloud is subject to diffusion and broadens with time. The width of the pulse visible in Figure 7.1, however, results not only from diffusion but also from a broadening effect by the preamplifiers which is non-negligible and must be subtracted. The complete analysis procedure for one set of 200 laser shots consists of five steps.

1. Fitting each of the 200 endpoints with a 2D Gaussian to obtain the longitudinal widths σ_t.

2. Determining $\overline{\sigma_t}$ from the distribution of σ_t.

3. Subtraction of the preamplifier broadening effect.

4. Modeling the longitudinal electric field E_L following the procedure in Chapter 4 to translate the width from time to spatial coordinates.

5. Calculating D_L by means of the drift time and Equation 1.21.

Concerning step 1, the fit model used to find the width of the endpoint signal is given by the two dimensional Gaussian

$$\frac{A}{2\pi\sigma_t\sigma_w} \exp\left[-\frac{\left(t-\bar{t}\right)^2}{2\sigma_t^2} - \frac{\left(w-\bar{w}\right)^2}{2\sigma_w^2}\right] + B, \qquad (7.2)$$

where A, B, σ_t, σ_w, $\bar{t}$ and $\bar{w}$ are free parameters describing the amplitude, a constant offset to account for the baseline, the longitudinal width (in units of ADC time samples), the transverse width (in units of wires) and the two coordinates of the peak location. An example is shown in Figure 7.6.

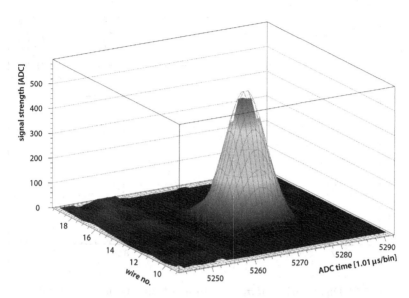

Figure 7.6: Example of an endpoint signal fitted with the model in Equation 7.2. The mesh represents the best fit.

To complete step 2, the values found for the parameter of interest σ_t for each of the 200 laser shots are accumulated in a histogram and the resulting distribution is fitted with a Gaussian to find the mean value $\overline{\sigma_t}$. One example of a distribution is depicted in Figure 7.7 along with a Gaussian fit.

To successfully decouple the effect of the preamplifier in step 3 and to obtain the effective longitudinal width $\sigma_{t,eff}$ of the charge cloud at the anode, resulting only from longitudinal diffusion, one must have knowledge of the preamplifier response to a δ input pulse. The shape of the response is Gaussian to a very good approximation (5th order shaper [109]) and has a width of $\sigma_{el} = 2.1 \pm 0.3\,\mu s$, determined from cosmic muons crossing the detector volume parallel to the wireplane and as close to it as possible. The latter requirement is necessary to make sure that there is no broadening effect of the charge cloud by diffusion. These ionization tracks produce input pulses to the preamp-

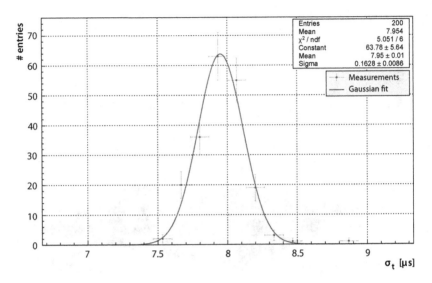

Figure 7.7: Distribution of σ_t as obtained from 200 laser shots along with a Gaussian fit. The fit yields $\overline{\sigma_t} = 7.95 \pm 0.01\,\mu s$.

lifier which are δ-like in the longitudinal direction and thus well suited to find the response width. Charge cloud broadening by diffusion also leads to a Gaussian distribution and hence the resulting signal after passing the amplifier is given by the convolution of two Gaussians of different widths $\sigma_{t,\text{eff}}$ and σ_{el}. To recover $\sigma_{t,\text{eff}}$, the recorded pulse (see Fig. 7.6) must be deconvoluted with the preamplifier response, meaning that

$$\sigma_{t,\text{eff}} = \sqrt{\overline{\sigma_t}^2 - \sigma_{\text{el}}^2}. \tag{7.3}$$

To translate the width $\sigma_{t,\text{eff}}$ from time to spatial coordinates $\sigma_{z,\text{eff}}$ in step 4, the longitudinal electric field E_L must be determined by means of the methods discussed in Chapter 4. Once E_L is known, one can relate the spatial coordinate z with the drift time coordinate via the field dependent drift velocity and hence transform $\sigma_{t,\text{eff}}$ to $\sigma_{z,\text{eff}}$. Since the Greinacher circuit is not always charged to the same level, even for identical voltage setpoints, a calibration of E_L must be done separately for each UV laser data set.

Finally, the longitudinal diffusion coefficient D_L is obtained using

$$D_L = \frac{\sigma_{z,\text{eff}}^2}{2t_d}, \tag{7.4}$$

where t_d is the total drift time resulting from the 2D Gaussian fit of the endpoint (parameter $\bar{t}$) after subtraction of t_0.

The procedure has been applied to 46 data sets recorded at 8 different Greinacher voltage setpoints ranging from 89 kV to 151 kV corresponding to average electric field strengths E_{avg} reaching from 120 ± 49 (RMS) V/cm to 210 ± 84 (RMS) V/cm, determined from the calibration procedure introduced in Chapter 4. The results are shown in Figure 7.8. To include the variations of the longitudinal electric field along the drift path, the RMS values of E_L are given below the data points. Measurement points sharing the same marker shape belong to the same series meaning that they were taken with identical laser settings (beam power, position and direction) and within a period of about 1 h. The error bars result from the statistical fluctuations as well as from the uncertainties in E_L and σ_{el}. The data points scatter relatively strongly in the range from $2.5\,\text{cm}^2/\text{s}$ to $6.6\,\text{cm}^2/\text{s}$. This cannot be explained by the statistical fluctuations or uncertainties resulting from the propagated errors. From what is known at present, it must be attributed to a systematic error originating from taking the data with different laser settings. These conclusions are drawn from Figure 7.8 as the measurements of one series are in agreement, while different series are shifted with respect to each other. By using the parameters obtained from the 2D Gaussian fits, it is possible to reveal the individual dependences of σ_t on the laser power and on the transverse width σ_w of the endpoint. It has been found that σ_t (and thus D_L) increases both with larger σ_w and with larger laser power. At least part of the latter dependence results from Coulomb repulsion effects. To obtain a full understanding of the values presented here, more advanced studies are necessary, though.

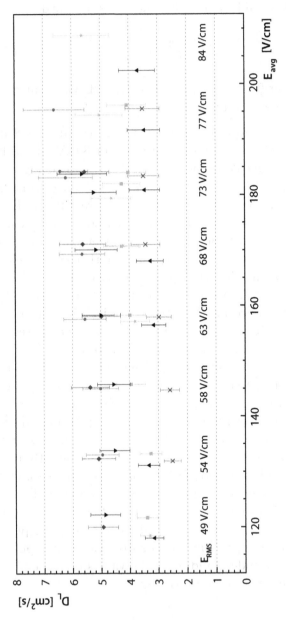

Figure 7.8: Overview on 46 individual measurements of D_L as a function of the electric field strength E_{avg} (mean values). RMS values of the electric field strengths at a given E_{avg} are indicated by the numbers below the data points. E_{avg} and E_{RMS} are obtained from the method introduced in Chapter 4 used to determine the course of E_L as a function of the drift coordinate. Data points sharing the same marker shape were measured with the same laser settings (beam power, position and direction).

When combining all the data points, one finds

$$D_L = \left(3.8 \pm 0.1 \,(\text{stat.}) \; ^{+1.7}_{-0.9} \,(\text{sys.})\right) \; \text{cm}^2/\text{s}, \qquad (7.5)$$

for electric field strengths ranging from $120 \pm 49\,(\text{RMS})\,\text{V/cm}$ to $210 \pm 84\,(\text{RMS})\,\text{V/cm}$. This value is consistent with the measurement of $4.8 \pm 0.2\,\text{cm}^2/\text{s}$ by [5], found in the same range of electric field strengths. A rough estimate for the diffusion coefficient can also be found by the following theoretical considerations [5]. The thermal energy in liquid argon amounts to about $kT_{BP} \approx 7.5\,\text{meV}$, where k is Boltzmann's constant and T_{BP} the boiling point of argon at a pressure of $1\,\text{atm}$ (see Tab. 1.1). At the mean electric field strength of about $165\,\text{V/cm}$, reached for the ARGONTUBE diffusion measurements, the electrons have drift velocities of the order of $0.8\,\text{mm}/\mu\text{s}$ [39], corresponding to a kinetic energy much lower than the thermal one. For thermal electrons, and using Equation 1.20, one can find a rough estimate for the expected theoretical longitudinal diffusion coefficient D_L^{th}. In the range of electric field strengths used for the measurement, the mobility of the electrons is given by $\mu = 458 \pm 59\,\text{cm}^2/\text{Vs}$ [39]. Plugging these values into Equation 1.20, one finds $D_L^{th} = 3.4 \pm 0.5\,\text{cm}^2/\text{s}$. However, this value does not take into account any effects resulting from the mutual Coulomb repulsion of the electrons within the cloud. Reference [5] quotes an approximate value for the contribution originating from the repulsion effect of $2\,\text{cm}^2/\text{s}$ for their measurement situation, to be added linearly to D_L^{th}.

Figure 7.9 shows a compilation of longitudinal diffusion measurements originating from [5] (square marker) and the ARGONTUBE (circles). The value given in Equation 7.5 is represented by the diamond marker (combined ARGONTUBE data) with statistical and systematic uncertainties added linearly. The solid line indicates the dependence of D_L on the electric field strength as given by the theory of Atrazhev and Timoshkin [106]. The horizontal error bars of the ARGONTUBE measurements correspond to the RMS values of the electric field strengths. Concerning the ICARUS and the combined-data ARGONTUBE measurement points, the horizontal errors are given by the full range of

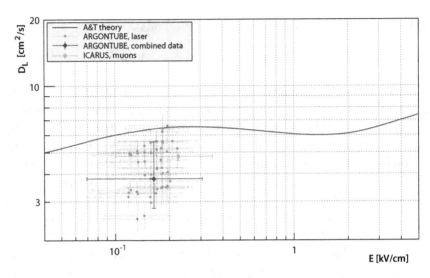

Figure 7.9: The longitudinal diffusion coefficient as a function of the electric field strength. The solid line marks the values for D_L expected from the theory by Atrazhev and Timoshkin [106]. The data points correspond to presently known measurements of D_L. Horizontal error bars for the ARGONTUBE data (circles) are given by E_{RMS}. For the combined ARGONTUBE data (diamond) as well as the ICARUS measurement (square) [5], they correspond to the full range of the electric field strengths in which the data were taken.

electric field strengths in which the measurements were performed. At least in the low electric field region ranging up to about 150 V/cm, the experimental data differ from the theoretical curve. The plot in Figure 7.9 clearly shows that more experimental data must be taken to allow a full understanding of the (longitudinal) diffusion processes in liquid argon in the presence of electric fields.

Conclusions

In the following, the contents of Chapters 3 to 7 are summarized and the main conclusions are highlighted again, including additional thoughts and ideas to further improve certain studies.

Based on descriptions in [42, 43, 76], I have designed and realized a regeneration system, allowing to recover the oxygen and water trapping capabilities of argon purifiers, and proved it to work by means of a residual gas analyzer (RGA). Before, the ARGONTUBE purifiers had to be regenerated by an external company. Having the possibility of doing this task in-house saves much time and costs. While testing and gaining experience with the system, I have worked out that the RGA is not a well suited device to monitor the moisture content in the gas stream during the regeneration procedure. It is a mass spectrometer and thus the identification of gas components is made difficult by the effects of fragmentation and multiple ionization. In addition, its reaction time is slow. Consequently, I have tested a second instrument, called a dewpoint meter, and the evaluation showed that it offers an unbiased, precise and fast measurement of the water content and hence serves as an ideal monitoring device for the given procedure.

An ingredient of major relevance to extract meaningful results from the cosmic muon and UV laser event samples that had been taken with the ARGONTUBE was to model and understand the electric field in the sensitive volume of the detector. UV laser induced tracks had proven that the field has disuniformities, mainly due to the Greinacher high voltage generator not reaching a fully charged state. The knowledge of the electric field strength in the sensitive detector volume is necessary to translate the drift time into the drift spatial coordinate z. The strength of a possible non-zero transverse component of the field would have to be known to also correct the x and y spatial coordinates. In addition, recombination losses which must be compensated for to obtain calorimetric information, depend on the electric field strength. I have used a model for the Greinacher circuit with only one free parameter to describe the electric potentials on the field cage rings in ARGONTUBE. From UV laser measurements, I was able to fix this parameter and to derive an analytic function describing the longitudinal electric field. By means of laser ionization tracks, corrected for the longitudinal field disuniformities, I have deduced an estimate for the parasitic transverse component of the electric field. The ratio between the transverse and the longitudinal field strengths was determined to be at most 1.6 %, given the restrictions in the accessible area of the sensitive volume with the current UV laser system. Considering the long drift distance and drift times in ARGONTUBE, even such a small transverse component can strongly distort the drifting ionization tracks as we have learned from our UV laser and cosmic muon induced events.

To process efficiently the large number of cosmic muon events that have been acquired with ARGONTUBE, I have implemented a software tool to identify and reconstruct the ionization tracks in the rawdata. To enable the separation of crossing tracks, I have employed a pattern recognition algorithm based on the Hough transform for straight lines. Several Hough space parameterizations [85] were taken into consideration to optimize the track separation capabilities and the computation efficiency. The PCLines algorithm offers advantages in both aspects [86, 87] and proved to be well suited for our purposes. I have achieved fast data processing by implementing compute-intensive parts of the code, such as the Hough transform, in the CUDA C language running on a NVIDIA graphics processing unit (GPU), a massively parallel computing architecture. By its nature, the Hough transform is highly qualified to be performed by a parallel computing architecture and I have shown that a significant speed-up compared to standard CPU serial code can be achieved. For an average ARGON-TUBE event, the PCLines Hough transform executed by a factor of

54.0 ± 0.5 faster on the GPU. The number of pixels containing signal is relatively low for ARGONTUBE events and the graphics card is not able to fully use its hardware resources for the small amount of data to be processed. The measurements proved that the hardware limit lies at a speed-up of 79.7 ± 0.1. Hence, exploiting the capabilities of GPUs for track reconstruction for various detector types, particularly for large LArTPCs, is highly recommended and has already been done by a number of experiments (see for instance [110–112]). Apart from the speed-up, the track finder turned out to be very helpful for the automatized analysis of the cosmic muon sample. In the future, its output could be used as an input, called a seed, for a reconstruction software based on the Kalman filtering technique.

I have taken a first look at the calorimetric capabilities of ARGONTUBE by studying the energy loss per unit length dE/dx for muons in liquid argon using two different read-out configurations. Mean values of

$$\langle dE/dx \rangle_{\text{warm}} = (2.23 \pm 0.02 \,(\text{stat.}) \pm 0.42 \,(\text{sys.})) \ \text{MeV/cm}, \quad (7.6)$$

$$\langle dE/dx \rangle_{\text{cryo}} = (1.82 \pm 0.01 \,(\text{stat.}) \pm 0.34 \,(\text{sys.})) \ \text{MeV/cm}, \quad (7.7)$$

and most probable values of

$$dE/dx_{\text{warm}}^{m.p.} = (1.56 \pm 0.27 \,(\text{sys.})) \ \text{MeV/cm}, \quad (7.8)$$

$$dE/dx_{\text{cryo}}^{m.p.} = (1.44 \pm 0.26 \,(\text{sys.})) \ \text{MeV/cm}, \quad (7.9)$$

were measured for respectively the former 'warm' and the new cryogenic read-outs. They are in agreement with literature values (see Tab. 1.1 and [101, 102]). For future analyses of the muon sample, the systematic errors can be greatly reduced when improving the processing of the bipolar induction signals [51]. This would not only increase the spatial resolution of the induction plane data, but also allow to obtain additional calorimetric information to increase the energy resolution. The implementation of full 3D tracking would improve the accuracy in measuring the inclination and position of tracks. Also, further investigations of the electric field disuniformities by means of the UV laser can lead to a better understanding of the transverse component of the field. Taking it into account would result in an improved spatial resolution for the x and y coordinates and thus in a higher accuracy in the track inclination and position measurements.

The charge to energy conversion factor used for the study of dE/dx for muons in liquid argon was obtained from preamplifier test bench measurements. A cross check of this value could be done by performing a self-calibration [101, 103] of the detector which makes use of the most probable energy loss per unit track length $dE/dx^{m.p.}$ for muons in liquid argon. This value depends on the momenta of the muons used for the analysis. If the momentum distribution was known, a Monte Carlo simulation could be performed to find the expected $dE/dx_{\mathrm{MC}}^{m.p.}$ in liquid argon and to determine the charge to energy conversion factor from the most probable amount of charge collected per unit track length $dQ/dx_{\mathrm{meas}}^{m.p.}$ obtained from the muon event sample. To find the momentum distribution of the acquired ARGONTUBE cosmic muon tracks, I have implemented a method for the momentum estimation on the basis of multiple Coulomb scattering effects for partially contained ionization tracks in LArTPCs. While the method itself has already been validated and proven to work [9], the implementation done for ARGONTUBE is yet to be verified, for instance by means of a Monte Carlo simulation. It can be further improved by using the Kalman filtering approach suggested in [9].

To extract calorimetric information from cosmic muon tracks, the amount of charge collected by the sensing wires of the TPC must amongst others be corrected for the electron attachment losses through electronegative impurities present in the detector volume. I have used two independent methods to measure the concentration of contaminant atoms and molecules in ARGONTUBE. In the first approach, a certain amount of charge was released from the TPC cathode by means of UV laser beam pulses (photoelectric effect). By measuring the amount of charge collected by the sensing wires as a function of the applied electric field strength, i.e. of the drift time, the concentration of oxygen equivalent impurities dissolved in the argon was obtained. Concerning the second approach, the most probable amount of collected charge per unit track length was evaluated as a function of the drift time for a large number of cosmic muon ionization tracks. The dependence was fitted with an exponential function to obtain the level of oxygen equivalent impurities in argon. The two methods were shown to give consistent results.

Finally, I have made a study of the longitudinal diffusion coefficient D_L in liquid argon at electric field strengths ranging from $120 \pm 49\,(\text{RMS})\,\text{V/cm}$ to $210 \pm 84\,(\text{RMS})\,\text{V/cm}$ with a final result of (combined data)

$$D_L = \left(3.8 \pm 0.1\,(\text{stat.})\,{}^{+1.7}_{-0.9}\,(\text{sys.})\right)\,\text{cm}^2/\text{s}. \qquad (7.10)$$

The value was obtained by shooting on the ARGONTUBE cathode with the UV laser and measuring the broadening of the released charge cloud after it had drifted across the full length of the TPC. The value is consistent with experimental results reported by ICARUS [5] at the given electric field strengths. The systematic errors in the measurement of D_L partly originate from the UV laser settings and need further investigation. Dependences of D_L on the laser power (Coulomb repulsion effect) and on the transverse beam width have been observed.

To summarize, the software tools that I have implemented and the methods that have been established within the scope of this master thesis will be relevant for future studies with ARGONTUBE. The measurement of the longitudinal diffusion coefficient in liquid argon marks an important contribution to future large mass and long drift distance LArTPCs.

Bibliography

[1] W. Pauli. Letter to L. Meitner about the neutrino hypothesis, 1930.

[2] E. Fermi. An attempt of a theory of beta radiation. *Z. Phys.*, 88:161–177, 1934.

[3] C. L. Cowan et al. Detection of the Free Neutrino: A Confirmation. *Science*, 124(3212):103–104, 1956.

[4] E.K. Akhmedov and A.Yu. Smirnov. Paradoxes of neutrino oscillations. *Phys. Atom. Nucl.*, 72:1363–1381, 2009.

[5] S. Amerio et al. Design, construction and tests of the ICARUS T600 detector. *Nuclear Instruments and Methods in Physics Research*, A527:329–410, 2004.

[6] C. Rudolf von Rohr et al. Argontube: an R&D Liquid Argon Time Projection Chamber. *Journal of Instrumentation*, 7(2), 2012.

[7] A. Ereditato et al. Design and operation of ARGONTUBE: a 5 m long drift liquid argon TPC. *Journal of Instrumentation*, 8 (7), 2013.

[8] M. Zeller. *Advances in liquid argon TPCs for particle detectors*. PhD thesis, University of Bern, 2013.

[9] A. Bueno et al. Measurement of Through-Going Particle Momentum By Means Of Multiple Scattering With The ICARUS T600 TPC. *The European Physical Journal C*, 48(2), 2006.

[10] J.N. Marx and D.R. Nygren. The Time Projection Chamber. *Physics Today*, 31(10):46–53, 1978.

[11] G. Charpak et al. The use of multiwire proportional counters to select and localize charged particles. *Nuclear Instruments and Methods*, 62(3):262–268, 1968.

[12] C. Rubbia. The liquid-argon time projection chamber: A new concept for neutrino detectors. *CERN-EP*, (77-8), May 1977.

[13] S. Klein. The time projection chamber turns 25. *Cern Courier*, 2004.

[14] S. Kubota et al. Dynamical behavior of free electrons in the recombination process in liquid argon, krypton, and xenon. *Phys. Rev. B*, 20(8):3486–3496, 1979.

[15] W. M. Haynes, editor. *Handbook of Chemistry and Physics*. CRC Press, 94th edition, 2013.

[16] C. Thorn. Properties of LAr. MicroBooNE Document Database, 2009.

[17] B. L. Henson. Mobility of Positive Ions in Liquefied Argon and Nitrogen. *Phys. Rev.*, 135:1002–1008, 1964.

[18] G. F. Knoll. *Radiation Detection and Measurement*. Wiley, 4th edition, 2010.

[19] E. Aprile et al. *Noble Gas Detectors*. Wiley, 2006.

[20] V. Chepel and H. Araújo. Liquid noble gas detectors for low energy particle physics. *Journal of Instrumentation*, 8(04), 2013.

[21] J. Beringer et al. (Particle Data Group). Review of Particle Physics. *Phys. Rev. D*, 86, 2012.

[22] M. A. Hofmann. *Liquid Scintillators and Liquefied Rare Gases for Particle Detectors*. PhD thesis, Technische Universität München, 2012.

[23] J. Sun, D. Cao, and J.O. Dimmock. Investigating laser-induced ionization of purified liquid argon in a time projection chamber. *Nuclear Instruments and Methods in Physics Research*, A370, 1996.

[24] B. Rossi et al. A prototype liquid Argon Time Projection Chamber for the study of UV laser multi-photonic ionization. *Journal of Instrumentation*, 7, 2009.

[25] I. Badhrees et al. Measurement of the two-photon absorption cross-section of liquid argon with a time projection chamber. *New Journal of Physics*, 12, 2010.

[26] L. Onsager. Initial Recombination of Ions. *Phys. Rev.*, 54: 554–557, 1938.

[27] G. Jaffe. The columnar theory of ionization. *Ann. Phys.*, 42, 1913.

[28] J. Thomas and D.A. Imel. Recombination of electron-ion pairs in liquid argon and liquid xenon. *Phys. Rev. A*, 36:614–616, 1987.

[29] U. Sowada et al. Hot-electron thermalization in solid and liquid argon, krypton and xenon. *Phys. Rev. B*, 25, 1982.

[30] B. Baller et al. A study of electron recombination using highly ionizing particles in the ArgoNeuT liquid argon TPC. *Journal of Instrumentation*, 8, 2013.

[31] S. Amoruso et al. Study of electron recombination in liquid argon with the ICARUS TPC. *Nuclear Instruments and Methods in Physics Research*, A523, 2004.

[32] J.B. Birks. *The Theory and Practice of Scintillation Counting*. Pergamon Press, 1964.

[33] A. Hitachi. Exciton kinetics in condensed rare gases. *J. Chem. Phys.*, 80, 1984.

[34] T. Heindl. *Die Szintillation von flüssigem Argon.* PhD thesis, Technische Universität München, 2011.

[35] T. Pollmann. *Pulse shape discrimination studies in a liquid argon scintillation detector.* PhD thesis, Max-Planck Institut für Kernphysik, 2007.

[36] S. Kubota et al. Recombination luminescence in liquid argon and in liquid xenon. *Phys. Rev. B*, 17(6):2762–2765, 1978.

[37] T. Doke. Fundamental Properties Of Liquid Argon, Krypton And Xenon As Radiation Detector Media. *Portugal Phys.*, 12 (9), 1981.

[38] L.S. Miller et al. Charge Transport in Solid and Liquid Ar, Kr, and Xe. *Phys. Rev.*, 166:871–878, 1968.

[39] W. Walkowiak. Drift Velocity of Free Electrons in Liquid Argon. *Nuclear Instruments and Methods in Physics Research*, A449: 288–294, 2000.

[40] K. Yoshino et al. Effect of molecular solutes on the electron drift velocity in liquid Ar, Kr, and Xe. *Phys. Rev. A*, 14:438–444, 1976.

[41] A.M. Kalinin et al. ATLAS-LARG-NO-058. ATLAS Internal Note, 1996.

[42] J. Spitz et al. A regenerable filter for liquid argon purification. *Nuclear Instruments and Methods in Physics Research*, A605: 306–311, 2009.

[43] G. Baptista et al. Experimental study on oxygen and water removal from gaseous streams for future gas systems in LHC detectors. Technical report, CERN, 2000.

[44] P. Cennini et al. Argon purification in the liquid phase. *Nuclear Instruments and Methods in Physics Research*, A333:567–570, 1993.

[45] K. Mavrokoridis et al. Argon purification studies and a novel liquid argon re-circulation system. *Journal of Instrumentation*, 6, 2011.

[46] C. Johnson. Results from the Fermilab Liquid Argon Purity Demonstrator. In *APS April Meeting Abstracts*, 2012.

[47] F. Bloch and N. E. Bradbury. On the mechanism of unimolecular electron capture. *Phys. Rev.*, 48(8):689–695, 1935.

[48] A. Herzenberg. Attachment of Slow Electrons to Oxygen Molecules. *The Journal of Chemical Physics*, 51(11), 1969.

[49] E. Buckley et al. A Study of Ionization Electrons Drifting over Large Distances in Liquid Argon. *Nuclear Instruments and Methods in Physics Research*, A275(2):364–372, 1989.

[50] J. Spitz. *Measuring Muon-Neutrino Charged-Current Differential Cross Sections with a Liquid Argon Time Projection Chamber*. PhD thesis, Yale University, 2011.

[51] C. Anderson et al. The ArgoNeuT Detector in the NuMI Low-Energy Beam Line at Fermilab. *Journal of Instrumentation*, 7, 2012.

[52] S. Navas-Concha et al. Analysis of the liquid argon purity in the ICARUS T600 TPC. *Nuclear Instruments and Methods in Physics Research*, A516:68–79, 2004.

[53] A. Bettini et al. A study of the factors affecting the electron lifetime in ultra-pure liquid argon. *Nuclear Instruments and Methods in Physics Research*, A305:177–186, 1991.

[54] G. Carugno et al. Electron lifetime detector for liquid argon. *Nuclear Instruments and Methods in Physics Research*, A292: 580–584, 1990.

[55] O. Bunemann, T. E. Cranshaw, and J. A. Harvey. Design of grid ionization chambers. *Canadian Journal of Research*, 27a (5):191–206, 1949.

[56] V. Radeka et al. Cold electronics for "Giant"Liquid Argon Time Projection Chambers. *Journal of Physics: Conference Series*, 308(1), 2011.

[57] D.W. Swan and T.J. Lewis. Influence of Electrode Surface Conditions on the Electrical Strength of Liquefied Gases. *J. Electrochem. Soc.*, 107(3):180–185, 1960.

[58] D.W. Swan and T.J. Lewis. The Influence of Cathode and Anode Surfaces on the Electric Strength of Liquid Argon. *Proceedings of the Physical Society*, 78(3), 1961.

[59] I. Kreslo et al. Experimental study of electric breakdowns in liquid argon at centimeter scale. *ArXiv e-prints*, 2014.

[60] F. Bay et al. Evidence of electric breakdown induced by bubbles in liquid argon. *ArXiv e-prints*, 2014.

[61] M. Auger et al. A method to suppress dielectric breakdowns in liquid argon ionization detectors for cathode to ground distances of several millimeters. *Journal of Instrumentation*, 9, 2014.

[62] C. Thorn et al. Cold Electronics Development for the LBNE LAr TPC. *Physics Procedia*, 37(0):1295–1302, 2012.

[63] H. Chen et al. Readout electronics for the MicroBooNE LAr TPC, with CMOS front end at 89K. *Journal of Instrumentation*, 7(12), 2012.

[64] G. De Geronimo et al. Front-End ASIC for a Liquid Argon TPC. *IEEE Transactions on Nuclear Science*, 58:1376–1385, 2011.

[65] M. Wutz, H. Adam, and W. Walcher. *Theorie und Praxis der Vakuumtechnik*. Vieweg, 2nd edition, 1982.

[66] Carbagas. Datasheet of argon, 2008.

[67] *Operation and Maintenance Manual for Surelite(TM) Lasers.* Continuum, 3150 Central Expressway, Santa Clara, CA 95051, USA, 2010.

[68] P. Lutz. G5601E Preamplifier Specification Sheet, May 2013. LHEP internal.

[69] Hamamatsu. Datasheet of Hamamatsu PMTs R7723, R7724, R7725, 2009.

[70] H.O. Meyer. Performance of a photomultiplier at liquid-helium temperatures. *Nuclear Instruments and Methods in Physics Research*, A621:437–442, 2010.

[71] R. Jerry et al. A study of the Fluorescence Response of Tetraphenyl-Butadiene. *ArXiv e-prints*, 2010.

[72] R. Acciarri et al. Aging Studies on thin tetra-phenyl butadiene films. *Journal of Instrumentation*, 8(10), 2013.

[73] A.V. Phelps and Z.Lj. Petrović. Cold-cathode discharges and breakdown in argon: surface and gas phase production of secondary electrons. *Plasma Sources Sci. Technol.*, 8, 1999.

[74] *RSV 151 / 301 / 601. RSV 151B / 301B / 601B. Roots Pumps.* Alcatel Hochvakuumtechnik GmbH, Am Kreuzeck 10 - Postfach 1151, 97877 Wertheim, Germany, 1995.

[75] *PHOENIXL 300. New Standards in Leak Testing.* Œrlikon Leybold Vacuum GmbH, Bonner Str. 498, 50968 Köln, Germany, 2007.

[76] R. Pisani. PHENIX Procedure for Regeneration of TEC-TRD Gas System Purifiers with Argon + 5% Hydrogen Mixture. Technical report, Brookhaven National Laboratory, 2004.

[77] H. Ibach. *Physics of Surfaces and Interfaces.* Springer, 2006.

[78] IUPAC. *Compendium of Chemical Terminology*. Blackwell Scientific Publications, 2nd edition, 2012.

[79] F.D. Gregory. Safety Standard for Hydrogen and Hydrogen Systems. Technical report, NASA, 1997.

[80] Air Liquide. Datasheet of ARCAL15, 2008.

[81] Shaw Moisture Meters. Superdew 3 Hygrometer. Operating Instructions.

[82] J.H. Gross. *Mass Spectrometry. A Textbook*. Springer, 2004.

[83] A. Zulkifli. *Polymeric Dielectric Materials*, chapter 1. Intech, 2012.

[84] S. Diaham and M-L. Locatelli. Dielectric properties of polyamide-imide. *Journal of Physics D: Applied Physics*, 46 (18):185302, 2013.

[85] A. Herout, M. Dubská, and J. Havel. *Real-Time Detection of Lines and Grids By PCLines and Other Approaches*. Springer, 2013.

[86] M. Dubská, A. Herout, and J. Havel. PCLines - Line Detection Using Parallel Coordinates. In *Proceedings of CVPR 2011*, pages 1489–1494. IEEE Computer Society, 2011.

[87] M. Dubská, A. Herout, and J. Havel. Real-Time Detection of Lines using Parallel Coordinates and OpenGL. In *Proceedings of SCCG 2011*. UNIBA, 2011.

[88] E.D. Church. LArSoft: A Software Package for Liquid Argon Time Proportional Drift Chambers, 2013. arXiv:1311.6774.

[89] P.V.C. Hough. Machine Analysis of Bubble Chamber Pictures. *Conf. Proc.*, pages 554–558, 1959.

[90] R.O. Duda and P.E. Hart. Use of the Hough transformation to detect lines and curves in pictures. *ACM*, 15:11–15, 1972.

[91] D.H. Ballard. Generalizing the Hough Transform to detect arbitrary shapes. *Pattern Recognition*, 13:111–122, 1981.

[92] R. Iser, B. Karimibabak, and S. Winkelbach. Hough transform java applet, 2005.

[93] J. Canny. A Computational Approach to Edge Detection. *IEEE Transactions on Pattern Analysis and Machine Intelligence*, PAMI-8(6):679–698, 1986.

[94] P. Bhattacharya, A. Rosenfeld, and I. Weiss. Point-to-line mappings as Hough transforms. *Pattern Recognition Letters*, 23: 1705–1710, 2002.

[95] V. Shapiro. Accuracy of the straight line Hough Transform: The non-voting approach. *Computer Vision and Image Understanding*, 103:1–21, 2006.

[96] C. Tu et al. A Super Resolution Algorithm to Improve the Hough Transform. In M. Kamel and A. Campilho, editors, *Image Analysis and Recognition*, volume 6753 of *Lecture Notes in Computer Science*, pages 80–89. Springer.

[97] S. Du et al. Measuring Straight Line Segments Using HT Butterflies. *PLoS ONE*, 7, 2012.

[98] NVIDIA. *CUDA C Programming Guide*. NVIDIA, 2013.

[99] J. Sanders and E. Kandrot. *CUDA by Example. An Introduction to General-Purpose GPU Programming*. Addison-Wesley, 2010.

[100] M. Harris. Optimizing CUDA, 2007. Slides.

[101] C. Anderson et al. Analysis of a Large Sample of Neutrino-Induced Muons with the ArgoNeuT Detector. *Journal of Instrumentation*, 7, 2012.

[102] O. Palamara et al. Performance of the $10\,m^3$ ICARUS liquid argon prototype. *Nuclear Instruments and Methods in Physics Research*, A498:292–311, 2003.

[103] R. Dolfini et al. Observation of long ionizing tracks with the ICARUS T600 first half-module. *Nuclear Instruments and Methods in Physics Research*, A508:287–294, 2003.

[104] C. Rudolf von Rohr. Monte Carlo simulation of cosmic muon interactions in the Argontube detector. Master's thesis, University of Bern, 2012.

[105] M.H. Cohen and J. Lekner. Theory of Hot Electrons in Gases, Liquids and Solids. *Phys. Rev.*, 158(2):305–309, 1967.

[106] V.M. Atrazhev and I.V. Timoshkin. Transport of Electrons in Atomic Liquids in High Electric Fields. *IEEE Transactions on Dielectrics and Electrical Insulation*, 5, 1998.

[107] E. Shibamura et al. Ratio of diffusion coefficient to mobility for electrons in liquid argon. *Phys. Rev. A*, 20(6), 1979.

[108] E.M. Gushchin et al. Electron dynamics in condensed argon and xenon. *Sov. Phys. JETP*, 55(4), 1982.

[109] Brookhaven National Laboratory. LARASIC4 (IC127) Datasheet, 2012.

[110] S. Gorbunov et al. ALICE HLT high speed tracking on GPU. *IEEE Trans. Nucl. Sci.*, 58(4):1845–1851, 2011.

[111] D. Emeliyanov and J. Howard. GPU-Based Tracking Algorithms for the ATLAS High-Level Trigger. *Journal of Physics: Conference Series*, 396(1), 2012.

[112] A. Ariga and T. Ariga. Fast 4π track reconstruction in nuclear emulsion detectors based on GPU technology. *ArXiv e-prints*, 2013.